THE ENNEAGRAM
OF G.I. GURDJIEFF

Mathematics, Metaphysics, Music, and Meaning

CHRISTIAN WERTENBAKER

Codhill Press
New Paltz, New York

CODHILL
Codhill Press books
are published for David Appelbaum

First Edition
Printed in the United States of America
Copyright © 2017 by Christian Wertenbaker

All rights reserved. No part of this publication protected by this
copyright notice may be reproduced or utilized in any form or
by any means, electronic, mechanical, including photocopying,
recording or by any informational storage and retrieval system,
without written permission of the publisher.

Cover image and design by Alicia Fox.
interior text design and illustrations by Alicia Fox
www.aliciafoxdesign.com

Library of Congress Cataloging-in-Publication Data

Names: Wertenbaker, Christian, 1943- author.
Title: The enneagram of G.I. Gurdjieff : mathematics, metaphysics, music, and
meaning / Christian Wertenbaker.
Description: first [edition]. | New Paltz, NY : Codhill Press, 2017. |
Includes bibliographical references and index.
Identifiers: LCCN 2017035379 | ISBN 1930337949 (alk. paper)
Subjects: LCSH: Gurdjieff, Georges Ivanovitch, 1872-1949. | Enneagram.
Classification: LCC BP605.G94 G879 2017 | DDC 197—dc23
LC record available at https://lccn.loc.gov/2017035379

THE ENNEAGRAM
OF G.I. GURDJIEFF

For Dolphi, Carlo, and Elena

Permissions

The author gratefully acknowledges the following rights holders for giving their permission to quote at some length:

Triangle Editions, Inc. from G.I.Gurdjieff,
Beelzebub's Tales to his Grandson, All and Everything, First Series

Tatiana Nagro, from P.D. Ouspensky
In Search of the Miraculous. Fragments of an Unknown Teaching

Table of contents

	Preface, major sources, acknowledgments	*xi*
1	Introduction	1
2	The basic mathematics of the enneagram	6
3	The meanings of symbols	12
4	One, two, three, and four	15
5	Five and ten	21
6	Six and symmetry	31
7	The unique properties of the number 7 and its relationships with the number 3	36
8	The law of seven and the seven-tone scale	45
9	Eight, nine, and the symmetry groups of the enneagram	61
10	The creation of the world and the three dimensions of time	67
11	The triangle and eternity—3, 6, and 9	74
12	More on the triangle, and the mingling of dimensions	76
13	The mirror symmetry of the enneagram	83
14	The four normed division algebras	87
15	Fermions and bosons—space and time	91
16	The inner and outer worlds, and the multiple intertwined aspects of the enneagram	93
17	The enneagram in movement(s)	96
18	More on the law of three	103
19	More on the law of seven	116
20	Conclusion	120
	Glossary	*125*
	References	*127*

Preface, major sources, acknowledgments

The enneagram is a symbol that was introduced to the modern world by G.I. Gurdjieff, who said that it had previously been kept secret and that it represented a complete description of the laws governing the universe. Because of the importance he attached to it, it has long intrigued followers of his teaching, and others, and has been put to various uses, but the understanding of its meanings remains very incomplete. This book is an attempt to further this understanding, which will nevertheless remain incomplete but perhaps a little less so. In particular, I have tried to relate the enneagram to a variety of mathematical and scientific ideas, since these also deal with the laws that govern the universe. As will be further elaborated below, Gurdjieff pointed out that symbols represent truths in a different way from scientific theories or mathematical formulas, and this has to be taken into account in trying to relate these different representations of reality. Nevertheless, there is only one reality, so it would seem useful to try to find connections between these different approaches, just as it seems useful to try to reconcile scientific and spiritual truths.

Several sources are quoted or referenced multiple times in the text and will be referred to by the following initials, followed by the page number for a particular quote or reference:

BT: Beelzebub's Tales to his Grandson. G.I. Gurdjieff, *All and Everything. First Series. Beelzebub's Tales to his Grandson*, (New York, Jeremy P. Tarcher/Penguin, 1992).

GET: Gurdjieff's Early Talks. *Gurdjieff's Early Talks* (Book Studio, 2014)

ISM: In Search of the Miraculous. P.D. Ouspensky, *In Search of the Miraculous: Fragments of an Unknown Teaching*. (New York: Harcourt, Brace & World, 1949).

MiC: Man in the Cosmos. Christian Wertenbaker, *Man in the Cosmos— G.I. Gurdjieff and Modern Science* (Codhill Press, 2012).

VRW: Views from the Real World. G.I. Gurdjieff, *Views from the Real World* (New York: E. P. Dutton & Co., Inc., 1973)

Other references, listed at the end of the book, will be denoted by the author's last name, or in the case of multiple authors, by the first author's.

I would like to thank the many people I have discussed these ideas with, who are too numerous to mention, but especially David Appelbaum, who has been my editor both at Codhill Press and Parabola magazine, for many conversations about Gurdjieff's ideas; Patricia Hemminger, Peter Kahan, Rick Sharpe, and Carl Nagin, for reviewing this and other manuscripts; Alicia Fox, for her expert illustrations and formatting of this book; and my wife Dolphi, for a long shared search in the Gurdjieff work.

I would also like to especially thank Stephen Grant, William Simon, and Tatiana Nagro for making it possible for me to quote extensively from P.D. Ouspensky and G.I. Gurdjieff's writings.

1
Introduction

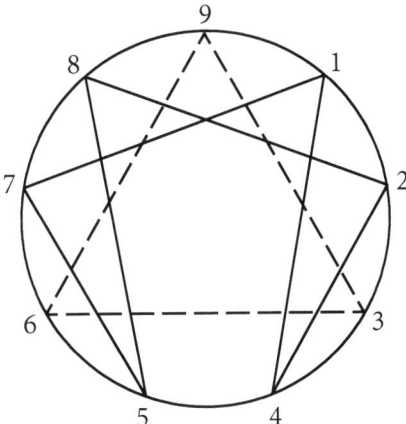

G.I. Gurdjieff considered the enneagram to be at the center of his teaching, even its emblem, a symbol in which one could read, if one knew how, all the aspects of the truths he had uncovered about the universe and humanity's place in it:

> …the enneagram is a *universal symbol*. All knowledge can be included in the enneagram and with the help of the enneagram it can be interpreted. And in this connection only what a man is able to put into the enneagram does he actually *know*, that is, understand. What he cannot put into the enneagram he does not understand. For the man who is able to make use of it, the enneagram makes books and libraries entirely unnecessary. *Everything* can be included and read in the enneagram. A man may be quite alone in the desert and he can trace the enneagram in the sand and in it read the eternal laws of the universe. And every time he can learn something new, something he did not know before.
> If two men who have been in different schools meet, they will draw the enneagram and with its help they will be able at once to establish which of them knows more and which, consequently, stands upon which step, that

is to say, which is the elder, which is the teacher and which the pupil. The enneagram is the fundamental hieroglyph of a universal language which has as many different meanings as there are levels of men.

The enneagram is *perpetual motion*, the same *perpetual motion* that men have sought since the remotest antiquity and could never find. And it is clear why they could not find *perpetual motion*. They sought outside themselves that which was within them; and they attempted to *construct* perpetual motion as a machine is constructed, whereas real perpetual motion is a part of another perpetual motion and cannot be created apart from it. The enneagram is a schematic diagram of *perpetual motion*, that is, of a machine of eternal movement. But of course it is necessary to know how to read this diagram. The understanding of this symbol and the ability to make use of it give man very great power. It is *perpetual motion* and it is also the *philosopher's stone* of the alchemists.

The knowledge of the enneagram has for a very long time been preserved in secret and if it now is, so to speak, made available to all, it is only in an incomplete and theoretical form of which nobody could make any practical use without instruction from a man who knows.

In order to understand the enneagram it must be thought of as in motion, as moving. A motionless enneagram is a dead symbol; the living symbol is in motion. (ISM, p. 294)

The enneagram is an integral part of many of Gurdjieff's movements, or sacred dances, and he said that "without taking part in these exercises, without occupying some kind of place in them, it was almost impossible to understand the enneagram." (ISM, p. 295)

It is possible to experience the enneagram by movement. The rhythm itself of these movements would suggest the necessary ideas and maintain the necessary tension; without them it is not possible to feel what is most important. (ISM, p. 295)

This clearly suggests that the intellect alone is not sufficient for understanding this symbol, nor, as Gurdjieff elaborates elsewhere, for truly understanding anything at all. Nevertheless, the intellect is a component of any understanding. Most of the explanatory material that Gurdjieff provided about the enneagram is in a few pages of P.D. Ouspensky's *In Search of the Miraculous* (ISM, pp. 286–295). There is also a record of a lecture by Gurdjieff on symbols and the enneagram in *Gurdjieff's Early Talks*, which Ouspensky seems to have made use of in his reporting. Otherwise there is little direct written material, and in particular, there is no mention of the enneagram *per se* in Gurdjieff's own writings, although there is much material on the laws of seven and three, which are expressed in the enneagram, and the law of seven is sometimes referred to as the "law of ninefoldness." Also, a case has been made that the actual structure of Gurdjieff's books reflects the enneagram and illuminates its meaning (Defouw).

The construction of the enneagram is based on certain properties of numbers, and an essential question which arises in any attempt to understand it is this: is the enneagram purely symbolic, in the sense that the geometric patterns created by these properties of numbers just happen to lend themselves to describing certain aspects of the world, or do the properties of numbers depicted in the enneagram actually underlie and determine the laws governing the universe. The same question can be asked about Gurdjieff's main laws of the universe, the law of seven and law of three. These are not explicitly the fundamental laws accepted by modern physics, although many parallels can be found between Gurdjieff's ideas and modern science. The most striking and basic parallel is that both in Gurdjieff's cosmology and in modern physics it is patterns of numbers that best describe both the fundamental constituents of the world and their interactions. Simple arrangements of whole numbers determine the configurations of electrons in atoms, and relatively simple matrices describe and predict the properties of subatomic particles. So, are the laws of seven and three merely symbolic of certain aspects of the world, or are they fundamental laws? If they are fundamental, their relationship to the currently accepted scientific laws remains to be clarified. There is ample evidence that Gurdjieff had great knowledge, so finding these relationships might be very illuminating. Furthermore, as will be elaborated in this book, the enneagram prominently includes both consciousness and time in its purview, two aspects of the world with which current scientific thought has great difficulty.

Gurdjieff himself was clearly of the opinion that the patterns represented in the enneagram are not just a convenient framework for symbolizing certain truths but are more basic. For instance, the patterns of the enneagram depend in part on the use of the decimal system, which many people regard as arbitrary, chosen perhaps because we have ten fingers. But Gurdjieff states that "the laws of unity are reflected in all phenomena. The decimal system is constructed on the basis of the same laws." (ISM, p. 289).

Most of the attempts to interpret and understand the enneagram, some of which are listed in the references, do not specifically address these questions, which require delving into the mathematics of the enneagram. A number of areas of mathematical inquiry appear possibly to be relevant to understanding the symbol, and this essay will concentrate on this aspect. It is hoped that those with a more extensive knowledge of mathematics and physics could take this further.

A brief overview of some of what will be proposed in this book might be helpful before plunging into details:

The enneagram depicts the laws of seven and of three, which Gurdjieff regarded as the fundamental laws governing processes in the universe, and particularly the processes by which the universe as a whole and each of its relatively autonomous parts—called cosmoses—maintain themselves. Gurdjieff regarded the universe as a living conscious being, made up of a hierarchy of cosmoses nested within each other, each also a living being with a certain level of consciousness. These cosmoses are somewhat variably described in his lectures and writings, but an acceptable list might be as follows:

God

The universe

The galaxy

The solar system

The planet

The multicellular organism—plants and animals

The cell or microbe

In addition, he regarded human beings, and similar beings elsewhere, as cosmoses distinct from other multicellular organisms. He called us tetartocosmoses, or three-brained beings, and regarded us as having a special role in the universe because of our capacity for the development of 'objective consciousness,' and for the creation of a soul, which can assume a supportive role in maintaining the universe's consciousness. The enneagram thus describes the possible evolution of man, as well as the evolution of the universe—this is the perpetual motion to which he refers—just as it describes other processes.

Processes follow a trajectory that, according to Gurdjieff, is described by the just-intonation major scale of Western music (this is the law of seven). This scale has two places where there is only a single semitone between adjacent notes (between other adjacent notes there are two semitones); these are places where help is needed for a process to continue without deviating. This help can come from outside or from within the organism itself. The three parts of the enneagram, the outer circle, the hexagram, and the triangle, interact to make this possible. In particular, these three parts represent three dimensions of time. In current physics, there is only one dimension of time, and the necessary consequence is that there must be a progressive overall increase in entropy, or disorder, in the universe. Gurdjieff thought that the consciousness of the universe is maintained, and this is possible by virtue of there being two other dimensions of time. Each of the dimensions of time corresponds to a particular brain of three-brained beings, or one could say that each brain is tuned to a different dimension of time, and this gives us our unique possibilities.

While there are recurring themes throughout this book, reflecting what I consider to be major aspects of the enneagram's meanings, there are also many loose ends and unresolved thoughts. Possibly significant connections between ideas have been included, even when the significance of these connections is unclear and may be nonexistent. One reason for including them anyway is that someone else may come along and find the missing links. The enneagram remains a great mystery.

2
The basic mathematics of the enneagram

In the decimal system, after 9 there is a return to 0, in a sense, as we arrive at the 10s; so at the apex of the enneagram, there is both 9 and 0, reflecting both a return to the starting point and also a jump to another level, just as in arithmetic. The numbers 0 through 9, 9 and 0 being the same point, are arranged in a circle on the enneagram. Similarly, in music, the octave of a note—twice the number of vibrations—is both the same note to our ears, and yet on a higher level. The notes of the scale are also represented around the circle of the enneagram, with do—both the lower do and the octave do—at position 9 (and 0). So the enneagram represents a spiral, depicted as a circle.

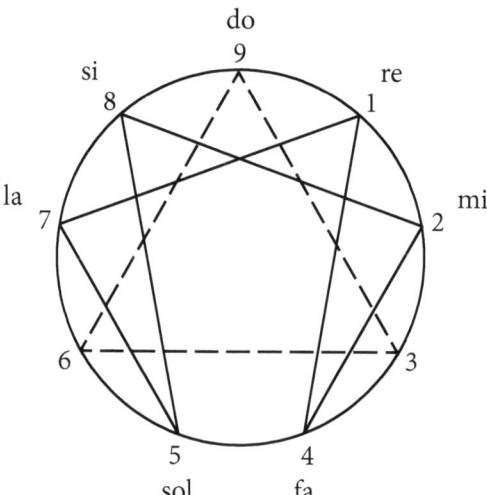

The points at 3 and 6 do not have a note there but represent the 'gaps,' or breaks, in the octave, the places where deviation of a process occurs, because of the missing semitone, unless outside help is provided at these locations to keep the process on track.

One can create a cyclic number system that only goes to 9 and then starts back at 0, which runs parallel to the decimal system.

A familiar example of this kind of cyclic number system is the clock, which starts over after 12. So 12 is also 0, and five hours after 8 o'clock, it is 1 o'clock. Other cyclical systems, like the days of the week, and the months of the year, behave similarly. In the case of a cyclical system based on 9, 9 is also 0, so 5 plus 5 is 1, 5 plus 6 is 2, etc. One can go back and forth between this cyclical system and the decimal system by "casting out nines," that is, subtracting nine repeatedly from a multi-digit number to arrive at a single digit, or equivalently, by adding the digits of the multi-digit number repeatedly to arrive at a single digit. So, 11 becomes 2 by either subtracting 9 or adding the two 1s. Similarly, 13 becomes 4, 23 becomes 5, and 99 becomes 0 by repeated subtraction of 9 or by adding the two 9s to make 18, and then adding 1 and 8 to make 9.

This is called arithmetic 'modulo 9' by mathematicians and is also known as 'theosophical addition,' the term which Gurdjieff used. Modular arithmetic is a well-developed branch of mathematics. Of course, there can be any modulus: in the case of the clock, the modulus is 12; in the case of the week, it is 7.

It makes sense that modular arithmetic might be applicable to the many physical phenomena that are cyclical in nature—vibrations of all sorts: atomic and molecular vibrations, sound vibrations, the electromagnetic oscillations of our brains, and the rotations and revolutions of celestial bodies.

One can use the relationship of the decimal system to modulo 9 arithmetic to check calculations. For instance:

$$283 \times 486 = 137538$$

To check the result, the two multiplicands are reduced to corresponding single digits by theosophical addition, i.e., to 4 and 9, the two digits are multiplied, to make 36, the single digit contraction of which is 9, and the result of the original multiplication should also reduce to 9, otherwise there is an error. In the days before calculators, this was a useful tool.

Regarding theosophical addition, Gurdjieff said:

> …for many people this method of the synthesizing of numbers appears so arbitrary that they regard it merely as a curious process, but one that is quite devoid of content. But all the while it has a deep significance for the man who has come to know the unity of existing things and who has the key thereto, reducing all multiform cycles to the basic facts which govern them. (GET, p.54)

In a sense, the checking of calculations in the decimal system by theosophical addition is a manifestation of this, and the importance in mathematics in general of modular arithmetic further attests to the validity of this statement.

On the other hand, the use of theosophical addition to reduce one's birthday to a single number, which is then regarded as significant in relation to one's character, is probably more in the "devoid of content" category.

The complicated six-sided figure (hexagram) in the enneagram, which connects the numbers 142857, comes about by dividing the numbers 1 through 6 by 7:

$$1/7 = 0.142857142857...\text{repeating forever}$$
$$2/7 = 0.285714285714...$$
$$3/7 = 0.428571428571...$$
$$4/7 = 0.571428571428...$$
$$5/7 = 0.714285714285...$$
$$6/7 = 0.857142857142...$$

The same digits 142857 occur in each fraction, in the same order, but starting in a different place.

$7/7$ is of course 1, or can be written as 0.9999999999... repeating forever.

The same pattern continues when the numerator of the fraction is larger than 7 but now with a number before the decimal point; thus, in a sense "jumping to a new level":

$$8/7 = 1.142857142857...$$
$$9/7 = 1.285714285714...$$
$$10/7 = 1.428571428571...$$
etc.

And:

$$15/7 = 2.142857142857...$$
$$16/7 = 2.285714285714...$$
$$22/7 = 3.142857142857...$$
$$23/7 = 3.285714285714...$$

And so on.

Among other geometric symmetries, the enneagram has bilateral, or mirror, symmetry across a vertical line through the center of the circle. There is a corresponding arithmetic symmetry in that the numbers on the same level on the two sides all add up to 9: 1 and 8, 2 and 7, 3 and 6, 5 and 4. Identically, if the first three numbers of the decimal expansion of $1/7$ are paired with the last three numbers, each pair adds up to 9, and the sum total is 999, which is 9 by theosophical addition:

```
  142
+ 857
-----
  999
```

Additionally, by theosophical addition, the six numbers of the sequence 142857 all add up to 9, which Gurdjieff says reflects the fact that an octave corresponds to a single note on another scale: octaves are nested one within another, like cosmoses.

> Making use of 'theosophical addition' and taking the sum of the numbers of the period, we obtain *nine,* that is, a whole octave. Again in each separate note there will be included a whole octave subject to the same laws as the first. (ISM, p. 289)

Even without further elaboration, the enneagram already demonstrates a remarkable pattern involving the numbers 1 through 9: the multiples of 3—3, 6, and 9—stand apart from the other digits, which form an interesting mirror-symmetric figure, derived from the decimal expansions resulting from division of the numbers 1 through 6 by 7. One can ask to what extent these patterns are unique or special, to what extent they depend on the decimal system, and what if anything is special about the decimal system itself in relation to other world phenomena. One can also ask what else is special about the number 7.

The decimal expansion of any fraction is either an exact number or a repeating decimal.

$1/2 = 0.5$
$1/3 = 0.3333....$
$1/4 = 0.25$
$1/5 = 0.2$
$1/6 = 0.16666....$
$1/7 = 0.142857142857142857....$
$1/8 = 0.125$
$1/9 = 0.1111....$

For the first 9 digits, the reciprocals of 2, 4, 5, and 8 produce exact results; those of 3, 6, and 9 produce single digit repeating decimals; and the pattern produced by $1/7$ is quite unique.

Going beyond 9:

$1/10 = 0.1$
$1/11 = 0.090909....$
$1/12 = 0.083333....$
$1/13 = 0.076923\ 076923....$ another complex repeating pattern, of 6 digits
$1/14 = 0.07142857142857....$ the same repeating pattern as $1/7$, as one might expect
$1/15 = 0.06666....$
$1/16 = 0.0625$
$1/17 = 0.0588235294117647\ 0588235294117647...$ another complex repeating pattern, with 16 digits
$1/18 = 0.05555....$
$1/19 = 0.052631578947368421\ 052631...$ an 18-digit repeating pattern
$1/20 = 0.05$
$1/21 = 0.047619\ 047619....$
$1/22 = 0.0454545....$
$1/23 = 0.0434782608695652173913\ 04347826....$ a 22-digit repeating pattern
$1/24 = 0.0416666....$
$1/25 = 0.04$
$1/26 = 0.0384615\ 384615....$
$1/27 = 0.0370370....$
$1/28 = 0.03571428\ 571428$
$1/29 = 0.0344827586206896551724137931\ 03448....$ a 28-digit repeating pattern

There are many interesting features of these numbers, long familiar to mathematicians. All fractions in decimal form are either exact, or produce repeating patterns, unlike numbers such as $\sqrt{2}$, or π, or e, or phi (about which more will be discussed later), which produce an endless series of digits without any apparent pattern. These numbers, however, are expressible as infinite series of fractions, with beautiful patterns of their own:

$$e = 1 + 1 + 1/2! + 1/3! + 1/4! + 1/5! ....$$
$$\pi/4 = 1 - 1/3 + 1/5 - 1/7 + 1/9 - 1/11 ....$$

! stands for factorial, which means that the number preceding the ! is multiplied by all the digits below it. $2! = 2 \times 1$, $3! = 3 \times 2 \times 1$, $4! = 4 \times 3 \times 2 \times 1$, etc.

It is logically necessary that the decimal expansion of a fraction produce either an exact number or a repeating pattern. If a number is divided by another, either there is no remainder, in which case the result is exact, or there is a remainder. When the remainder times 10 is then divided by the denominator, to arrive at the next decimal place, the same two possibilities obtain. Eventually, one either arrives at no remainder, in which case the result is exact, or at a remainder that is the same as one of the previous remainders, in which case the digits of the result will repeat themselves endlessly.

Other interesting features of these fractions are as follows: 1) the complex repeating patterns often have a number of digits in the pattern that is one less than the original divisor; 2) typically the sequence can be divided in two: each digit in order from the start of the sequence is the complement (adding up to 9) of each digit in order beginning from the middle.

In any case, the pattern formed by the fraction $\frac{1}{7}$ is unique among the first 9 digits, and although it is only the first of a series of complex repeating patterns of digits produced by fractions, it is the most striking, concise, and beautiful in its inclusion of all the digits except for 3, 6, and 9, each only once.

Of course, fractions in any other number system will obey the same rules, and there will be instances of repeating patterns formed by divisional expansions in those number systems. For instance, in a number system based on 12, $\frac{1}{7} = 1\,8\,6\,10\,3\,5\,1\,8\,6\,10\,3\,5\,\ldots$ forms an enneagram-like geometric pattern if the numbers 1 through 11 are placed around a circle and the numbers of the repeating numerical pattern are connected in order. Additionally, there are related symmetries: the first three and second three numbers arranged in pairs add up to 11.

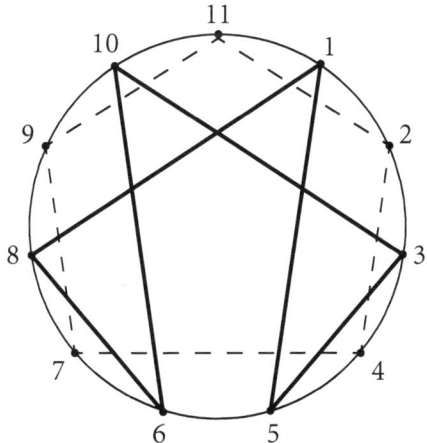

3

The meanings of symbols

For Gurdjieff, as for Pythagoras, who was one of the few historical figures that Gurdjieff greatly admired, numbers, particularly the first 10, were not only tools for calculation but also meaningful symbols related to fundamental truths about the organization of the world. Numbers also have geometric representations, which are equivalent symbols. So, 2 can be represented by an equal sign or just by a line between two points, 3 by a triangle, 4 by a cross or square, 5 by a pentagon or five-pointed star, 6 by a hexagon or six-pointed star, etc.

The enneagram is a more complex symbol, with several components. Gurdjieff regarded symbols as multifaceted in their representations and held that the multiple meanings attached to a symbol gave it a different kind of exactitude, compared to an ordinary definition or a logical sequence.

> ...the intellectualism of contemporary education imbues people with a propensity and a tendency to look for logical definitions and for logical arguments against everything they hear and, without noticing it, people unconsciously fetter themselves with their desire, as it were, for exactitude in those spheres where exact definitions, by their very nature, imply inexactitude in meaning.... This does not mean that exact definitions do not exist on the way of true knowledge, on the contrary, only there do they exist; but they differ very greatly from what we usually think them to be. (ISM, p. 284)

Gurdjieff regarded symbols as reflecting what he called 'objective knowledge,' as opposed to ordinary scientific observations and theories, which resulted in 'subjective knowledge.' The capacity for objective knowledge resided in what he called the 'higher intellectual center,' capable of thought that greatly transcends ordinary thought. One clue as to what this means is in a definition he gave of true consciousness:

> *Consciousness* is a state in which a man *knows all at once* everything that he in general knows... (ISM, p. 155)

One gets a sense of the direction Gurdjieff is indicating when suddenly, in an instant, one comprehends a large multifaceted set of ideas, which at other times required plodding through, one association at a time; and this insight subsequently takes many hours of explanation to convey in words. It is said that Mozart could hear an entire symphony in his mind all at once, and there are many reports of scientists who, after struggling with a set of ideas for a long time, suddenly, in a flash, understand all the relations between them.

Descriptions of states of 'enlightenment' point in the same direction:

> In this ultimate state of open availability, the mind instantaneously grasps everything. (Bouanchaud, p. 157)

What is the nature of the change in thought that makes this possible? One of Gurdjieff's main ideas is that knowledge is the province not only of the mind but also of the emotional and physical aspects of the nervous system—that we are 'three-brained beings.'

> …we must understand that every normal psychic function is a means or an instrument of knowledge. With the help of the mind we see one aspect of things and events, with the help of emotions another aspect, with the help of sensations a third aspect. The most complete knowledge of a given subject possible for us can only be obtained if we examine it simultaneously with our mind, feelings, and sensations.… In ordinary conditions man sees the world through a crooked, uneven window. (ISM, p. 107–108)

This three-brained perception is of a higher dimensionality than ordinary logical thought. Just as visual depth perception requires a fusion in the brain of the slightly different views from the two eyes, objective knowledge requires a fusion of the inputs from three different organs of perception ("Shadows of the Real World." In MiC, pp. 65–75). This additional dimensionality provides a precision that is not possible otherwise. An object precisely located in three-dimensional space can be in different locations in different two-dimensional shadows of this space. Similarly, the true understanding of a symbol depends on a fusion of its many associated meanings, which are not just mental.

> …a symbol…possesses an endless number of aspects from which it can be examined and it demands from a man approaching it the ability to see it simultaneously from different points of view. (ISM, p. 283)

With this in mind, one can examine the meanings attached to the first 10 numbers, by both Pythagoras and Gurdjieff, who taught quite similarly in this respect, and then relate these meanings to the mathematical properties of the numbers. Just as the enneagram is the emblem of Gurdjieff's teaching, the emblem of the Pythagoreans was the tetractys, a representation of the numbers 1–10:

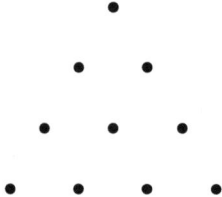

It is not difficult, nor does it seems arbitrary, to superimpose the enneagram on the tetractys:

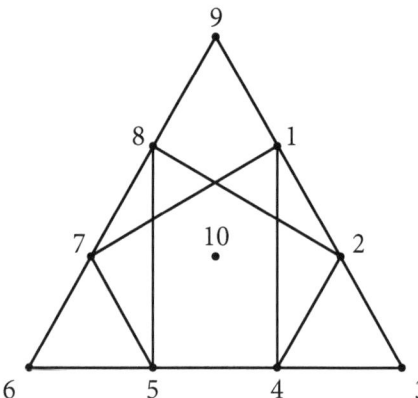

4
One, two, three, and four

It is fairly straightforward to see how the numbers 1, 2, and 3 represent fundamental aspects of the world we live in. One represents the separate existence of each defined thing, organism, or concept, and also the unity of the whole. According to Gurdjieff:

> One of the most central of the ideas of objective knowledge is the idea of the unity of everything, of unity in diversity. (ISM, p. 278)

> The monad, manifest as the number one, denotes the primordial unity at the basis of creation. (Strohmeier, p. 70)

The number 2 represents opposites, conflict, and interaction. Opposites abound in the world: even and odd numbers, positive and negative charge, male and female, attraction and repulsion, etc. The interaction of opposites can continue as conflict or can produce a result, or a reconciliation, represented by the number 3. All equations denoting mathematical, physical, chemical, or any other interactions, contain at least three elements; otherwise, they are just definitions. Thus we have Newton's second law, force = mass × acceleration, Ohm's law of electrical circuits, voltage = current × resistance, etc. An equation such as x = y is only stating an identity. This will be elaborated further in chapter 18. Gurdjieff considered the 'law of three' to be one of the fundamental laws of the universe. He called the three elements active, passive, and neutralizing and considered all phenomena to be the result of three forces. Although we tend to see two forces as producing an interaction, or a result, he regarded this as a limited view:

> According to real, exact knowledge, one force, or two forces, can never produce a phenomenon. The presence of a third force is necessary, for it is only with the help of a third force that the first two can produce what may be called a phenomenon, no matter in what sphere.
>
> The teaching of the three forces is at the root of all ancient systems. The first force may be called active or positive; the second, passive or negative; the third, neutralizing. But these are *merely names*, for in reality all three forces are equally active and appear as active, passive, and neutralizing,

only at their meeting points, that is to say, *only in relation to one another at a given moment*. The first two forces are more or less comprehensible to man and the third may sometimes be discovered either at the point of application of the forces, or in the 'medium,' or in the 'result.' But, speaking in general, the third force is not easily accessible to direct observation and understanding. The reason for this is to be found in the functional limitations of man's ordinary psychological activity and in the fundamental categories of our perception of the phenomenal world, that is, in our sensation of space and time resulting from these limitations. People cannot perceive and observe the third force directly any more than they can spatially perceive the 'fourth dimension.' (ISM, p. 77)

What does this mean? The statement that the third force can "sometimes be discovered…in the 'result'" may provide a clue. We usually think in terms of a causal and temporal chain going from causes to results and from past to future, but the result can also be seen as an ingredient in an interaction acting at the same time as the other causes, in that it can be an intention in the mind of a conscious being. This reflects the fact that consciousness adds another dimension to time, as will be further discussed below.

That the third force can sometimes be discovered in the medium reflects the fact that the medium can be of a different, typically finer, materiality than the two opposing forces. Thus the three forces may come from different levels, or cosmoses, as also further discussed below. In the example above, the intention of a conscious being can be on a different level than the other two forces.

There are any number of fundamental triads to be found in the world: the three dimensions of space we live in; past, present, and future; the three fundamental particles composing atoms (proton, neutron, and electron), and further, the three classes of leptons (electron, muon, and tau), the three 'colors' and three classes of 'flavors' of quarks. It takes three lines to make an enclosed figure in the plane, the triangle. We have three modes of perception, corresponding to our three brains. Every process has a beginning, middle, and end. There are three basic grammatical subjects, corresponding to three basic relationships: I, you, and (s)he, or we, you, and they. In religious teachings, three figures prominently: the Father, Son, and Holy Ghost; Brahma, Vishnu, and Shiva.

In the Samkhya school of Hindu philosophy, there are three *gunas*—qualities, or tendencies—*rajas, tamas,* and *sattva*. *Rajas* is the quality of activity and passion; *tamas* of inertia, apathy, and dullness; and *sattva* of balance, knowledge, and harmony. All three qualities are present in any person, in variable proportions. These proportions are detectable in the quality of the pulse. But these are not just human qualities but also universal principles that apply to everything, much like Gurdjieff's three forces.

> The three gunas—*sattva*, or illumination and truth, *rajas*, or passion and desire, and *tamas*, or darkness, sloth and dullness—were originally thought, by the Samkhya philosophers who first identified and named them, to be substances. Later they became attributes of the psyche. *Sattva* has been equated with essence, *rajas* with energy and *tamas* with mass. According to still another interpretation, *sattva* is intelligence, *rajas* is movement and *tamas* is obstruction. The word *guna* means 'strand', 'thread', or 'rope' and *prakrti*, or material nature is conceived as a cord woven from the three *gunas*. (Sargeant, *Bhagavad Gita,* note on book VII, verse 13)

These interpretations of the three forces are not far from Gurdjieff's, nor from the scientific division of all aspects of the world into matter, energy, and information.

There is even the idea, in Samkhya philosophy, of the creation of the material world being the result of an imbalance of the gunas, reminiscent of Gurdjieff's idea that the creation resulted from a change in the symmetry of the fundamental laws governing the universe, which will be further discussed later.

> Before creation, they [the gunas] remain inactive and in a state of perfect balance in the Primordial Nature (mula Prakriti). When their balance is disturbed, creation sets in motion, and the diversity of objects and beings comes into existence, each possessing the triple gunas in different proportions. (Jayaram)

Four may at first seem a bit less fundamental, and its features blend with those of three. Gurdjieff almost skips over it in his discussion of symbols in ISM. We divide the year into four seasons and the surface of the earth into four directions, but is this arbitrary? Perhaps not—a line with two directions and another line perpendicular to it are fundamental aspects of the construction of the world, are required for the depiction of the complex numbers (chapter 14), and are seen in the basic distinction between interactions on one level and those involving another (horizontal: same level; vertical: another level). The cross symbolizes four.

A light ray, as depicted by physics, consists of perpendicular electric and magnetic field oscillations, both of which are perpendicular to the direction of motion. While the magnetic and electric fields are equal in this situation, they represent different levels at speeds slower than lightspeed, because an electric field exists around a charge that is not moving, but a magnetic field only around a moving charge. For that matter, light and other electromagnetic waves exist in a different dimensionality from the ordinary three-dimensional space we are used to, as evidenced by their invariant velocity regardless of the velocity of the source ("The Light of the Beholder." In MiC, pp. 49–56). So, perpendicular electric and magnetic field oscillations represent different levels.

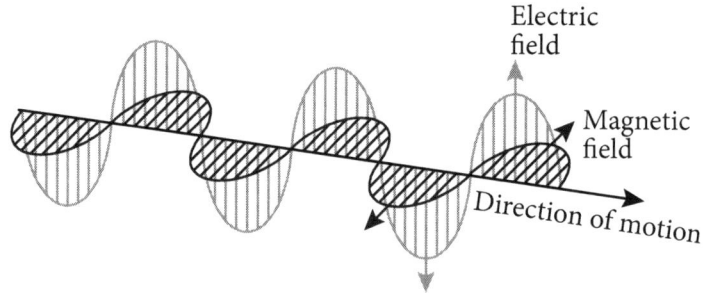

Gurdjieff regarded all phenomena as requiring the interaction of three forces, not just two, and as described above, considered the third force as more difficult to perceive. He suggests in the quote above that this is a question of dimensionality, and this can be interpreted to mean that the third force is in a different dimension, or on a different level. The relationship between levels, dimensions, and cosmoses, already alluded to, is all important in Gurdjieff's ideas and will be discussed throughout this book. He also considered electromagnetism, which he called 'okidanokh' to consist of three forces, which he called 'anodnatius,' 'cathodnatius,' and 'parijrahatnatius' (BT, pp. 146–147). The first two names clearly refer to positive and negative charges, or poles, but what is the third force in electromagnetism? Perhaps the magnetic field itself. The magnetic field produced by a current between the two opposite poles in a sense binds and reconciles them.

In an obscure chapter in *Beelzebub's Tales* ("The Arch Preposterous." In BT, pp. 140–165), Gurdjieff describes experiments in which different proportions of the three components of okidanokh, directed at various chemical elements, would transform them into other elements. While this is not a known property of the electromagnetic force, the weak nuclear force *is* the one that governs radioactive decay, whereby elements do transform one into another, and the weak nuclear force and the electromagnetic force are considered by modern physics to be related: the two forces are one at higher energies but separate at lower energies because of a broken symmetry. Did Gurdjieff anticipate this discovery as he did a number of others, such as the genesis of the moon and the existence of multiple galaxies?

At least four points are needed to define a geometric solid, the tetrahedron being the simplest one. Four also completes the tetractys, 1, 2, 3, and 4 adding up to 10. The tetractys was regarded by the Pythagoreans as symbolic of the cosmos in many ways. Geometrically, it represents the dimensions of space: the point; the line, defined by two points; the plane, represented by the triangle; and the solid, the tetrahedron, defined by four points. It also represents the emergence of multiplicity from unity, culminating in the number 10, which denotes in the decimal system a return to unity, now on a higher level (Guthrie). On the way from unity to unity, one passes by opposition and conflict, represented by two, and reconciliation and harmony, represented by three. Four represents a solidification of this process.

Gurdjieff discussed these numbers in relation to man's development:

> Man, in the normal state natural to him, is taken as a *duality*. He consists entirely of dualities or 'pairs of opposites.' All man's sensations, impressions, feelings, thoughts, are divided into positive and negative, useful and harmful, necessary and unnecessary, good and bad, pleasant and unpleasant. The work of centers proceeds under the sign of this division. Thoughts oppose feelings. Moving impulses oppose instinctive craving for quiet. This is the duality in which proceed all the perceptions, all the reactions, the whole life of man. Any man who observes himself, however little, can see this duality in himself.
>
> But this duality would seem to alternate; what is victor today is the vanquished tomorrow; what guides us today becomes secondary and subordinate tomorrow. And everything is equally mechanical, equally independent of will, and leads equally to no aim of any kind. The understanding of duality in oneself begins with the realization of mechanicalness and the realization of the difference between what is mechanical and what is conscious. This understanding must be preceded by the destruction of the self-deceit in which a man lives who considers even his most mechanical actions to be volitional and conscious and himself to be single and whole.
>
> When self-deceit is destroyed and a man begins to see the difference between the mechanical and the conscious in himself, there begins a struggle for the realization of consciousness in life and for the subordination of the mechanical to the conscious. For this purpose a man begins with endeavors to set a definite *decision*, coming from conscious motives, against

mechanical processes proceeding according to the laws of duality. The creation of a permanent third principle is for man the *transformation of the duality into the trinity*.

Strengthening this decision and bringing it constantly and infallibly into all those events where formerly accidental neutralizing 'shocks' used to act and give accidental results, gives a permanent line of results in time and is the *transformation of trinity into quaternity*. (ISM, p. 281–282)

There is a striking parallel between the above description of "the creation of a permanent third principle" in man, and the idea that the magnetic field is the third force in electromagnetism. Gurdjieff speaks of 'animal magnetism' as the "blood" of the second body, or first stage of the soul (BT, pp. 519–520, 666), and its production comes about through the struggle between positive and negative in man (BT, p. 1100).

Four is also the number of the quaternions, a system of complex numbers with one real part and three imaginary parts, which is important in modern physics as it provides a mathematical scaffold for the weak nuclear force, and about which more will be discussed in chapter 14.

5
Five and ten

The numbers 5 and 10 appear in many places, and this ubiquity may help justify the apparent primacy of the decimal number system.

Aside from our fingers and toes, the numbers 5 and 10 appear in nature, such as in the pentagonal symmetry of certain flowers and starfish; the DNA molecule, seen on end, forms a ten-sided figure.

Two of the *five* regular platonic solids, the icosahedron and the dodecahedron, have pentagonal elements.

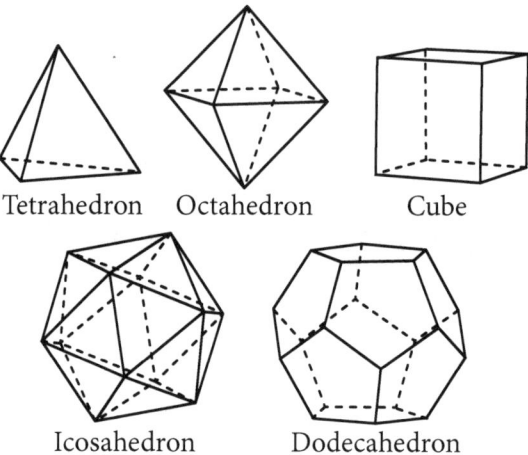

Tetrahedron   Octahedron   Cube

Icosahedron   Dodecahedron

One of the most intriguing places where the number 5 turns up repeatedly is in connection with the golden mean, or phi (Ø). The golden mean, or golden section, is the ratio that results when a line is divided into two segments, such that the ratio of the length of the whole line to the length of the longer segment is the same as the ratio of the length of the longer segment to the length of the shorter segment.

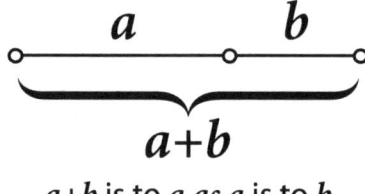

a+b is to *a* as *a* is to *b*

The formula for this ratio, for a shorter segment set equal to 1, i.e., b =1, is (a+1)/a=a/1, or $a^2=a+1$, or $a^2-a-1=0$. Solving, (using the quadratic formula), produces: a=$(1+\sqrt{5})/2$, called phi (Ø), which is an irrational number equal to approximately 1.618, or $(1-\sqrt{5})/2$, called phi prime(Ø'), approximately equal to 0.618.

The golden section has been a source of fascination since ancient times, appearing in geometry (notably in the pentagonal star, another symbol of the Pythagoreans), in art, and in nature in a variety of places.

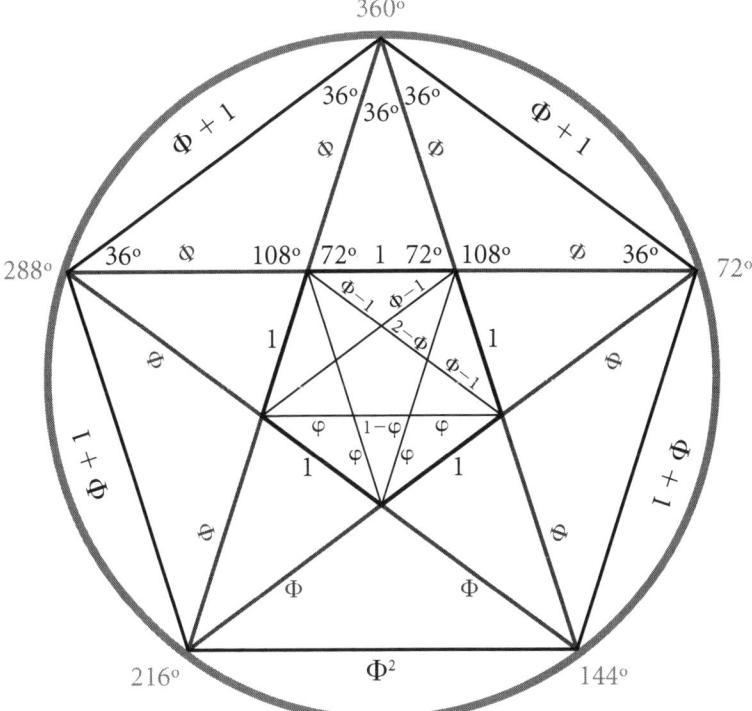

The ratio of the longer to the shorter side of each triangle is Ø, as is the ratio of the longer plus the shorter side to the longer side.

By repeatedly applying this ratio (each of the rectangles in the image below is a 'golden rectangle'—the ratio of the longer to the shorter side is Ø), self-similar growing patterns are formed:

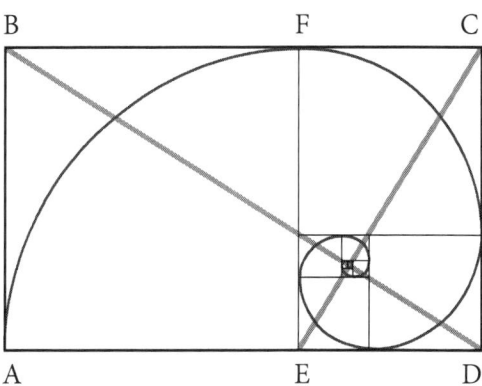

The spiral generated in this way, called the logarithmic spiral, is found in certain shells, notably nautilus:

Thus the golden ratio is a means for growing organisms to retain the same shape as they grow, which is perhaps one reason it is so ubiquitous.

The golden ratio, phi, has a number of intriguing and unique mathematical properties:

$$\emptyset^2 = 1+\emptyset$$
$$1/\emptyset = \emptyset-1$$
$$\emptyset-1 = \emptyset'$$

It is also unique in its having the simplest continued fraction. Continued fractions were known to the ancient Greeks. When a number is divided by a smaller number, if there is a remainder it can be inverted, as shown below, and then the process can be continued until it stops when there is no remainder:

$$\frac{74}{32} = 2 + \frac{10}{32} = 2 + \frac{1}{\frac{32}{10}} = 2 + \frac{1}{3+\frac{2}{10}} = 2 + \frac{1}{3+\frac{1}{\frac{10}{2}}} = 2 + \frac{1}{3+\frac{1}{5}}$$

The continued fractions of rational numbers stop, but those of irrational numbers go on forever, as do their decimal expansions. In the case of phi, the continued fraction goes on forever with just 1s, making it the simplest irrational number, and, in a sense, the "most irrational" of numbers (Baez).

$$\emptyset = 1 + \cfrac{1}{1 + \cfrac{1}{1 + \cfrac{1}{1 + \cfrac{1}{1 + \cfrac{1}{1 + \ldots}}}}}$$

Phi is related to pi, by the following formula:

$$\emptyset = 2 \cos(\pi/5)$$

Intimately related to the golden mean is the Fibonacci series of numbers, formed by starting with 0 and 1, and then adding adjacent digits to produce the next:

$0+1=1, 1+1=2, 2+1=3, 3+2=5, 5+3=8$, etc.

giving the sequence:

0, 1, 1, 2, 3, 5, 8, 13, 21, 34, 55, 89, 144, 233, etc.

The ratios of adjacent Fibonacci numbers converge on phi:
1:1=1, 2:1=2, 3:2=1.5, 5:3=1.666, 8:5=1.6, 13:8=1.625, 21:13=1.6153846, 34:21=1.6190476, 55:34=1.617647, 89:55=1.6181818, 144:89=1.6179775, 233:144=1.6180555. $\emptyset$=1.6180339887…

The Fibonacci numbers are also found in nature in a variety of places, such as in the opposing spiral patterns of sunflowers and some cactuses:

Counting the spirals in one direction and those in the other direction, one finds two adjacent Fibonacci numbers.

The Fibonacci numbers also appear in the branching patterns formed by following certain rules, which apply to the genealogy of drone bees, for instance, and to the pattern of branches of a sneezewort (Huntley, pp. 160–163).

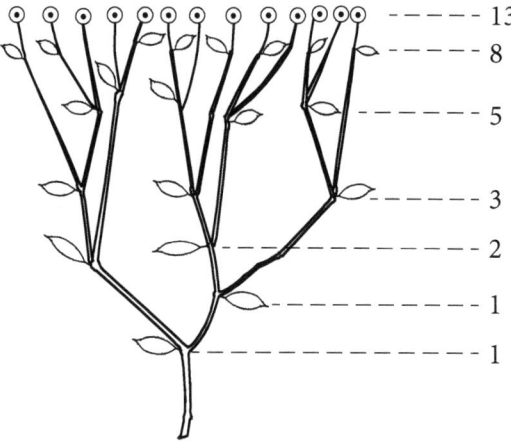

The rules are that any given branch can divide in two either after one generation or two, and one of the two branches of a division will divide again after one generation, and the other after two.

Fibonacci numbers and the golden ratio are also intimately related to another area of mathematics, fractals. Fractals are structures that are self-similar on multiple scales, as in the case of the Sierpinski triangle:

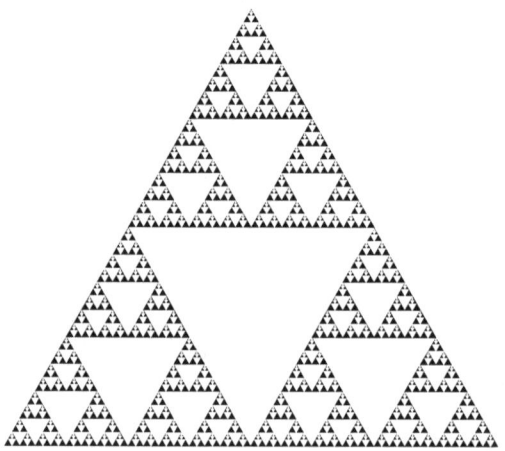

Many natural phenomena such as coastlines, mountains, and clouds show self-similarity, which means that they look much the same and have the same patterns of irregularities, up to a point, when seen from different distances, on different scales. The same is true of some aspects of living organisms: the branching of trees repeats itself on multiple scales in a self-similar fashion. Since the golden ratio is also a mechanism for growth without change of shape, it makes sense that these two areas are related.

One can play the following game to illustrate this. Start with 1 and create a series of numbers by following two rules: a 1 changes to a 10 and a 0 to a 1:

NUMBER	NUMBER OF DIGITS
1	1 (one 1)
10	2 (one 1, one 0)
101	3 (two 1s, one 0)
10110	5 (three 1s, two 0s)
10110101	8 (five 1s, three 0s)
1011010110110	13 (eight 1s, five 0s)
101101011011010110101	21 (13 1s, eight 0s)

The number of digits of each number in the sequence follows a Fibonacci sequence. Further, each number can be made by stringing together the two above it, and the numbers of 1s and 0s also follow Fibonacci sequences. These patterns are also self-similar on different scales. But they are not crystalline like the Sierpinsky triangle: there is long range order but not total regularity. Like five-sided geometric figures tiling the plane, they represent a mixture of order and randomness (see p. 31).

The vascular and nervous systems have a tree-like structure, and are self-similar on different scales. It is required that they penetrate everywhere in the body to fulfill their functions. It turns out that the ideal ratio of the lengths of a daughter branch to a mother branch, so that a tree will fill space, is $\emptyset'$, 0.618 (Livio).

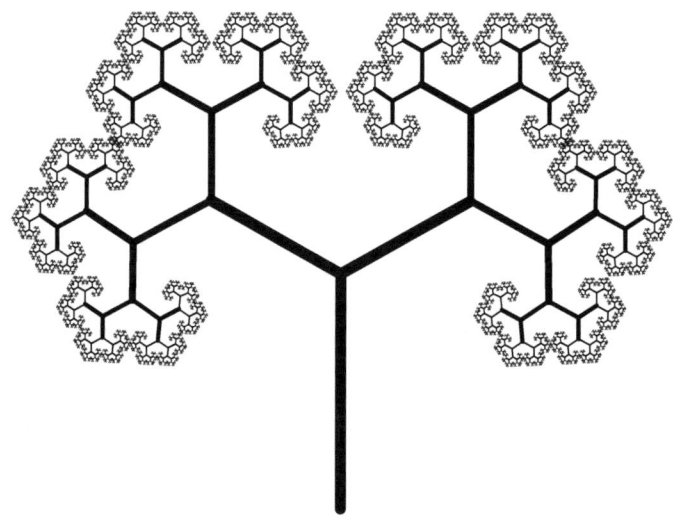

'Golden Tree,' with Ø¹ branching ratio

* * *

The number 10, aside from being twice 5, also appears in interesting and odd places in the world of mathematics and physics. The tetractys of course has ten points, and as already mentioned, the DNA molecule, seen on end, has ten sided symmetry. There are ten components of Riemann's metric tensor, an essential part of the mathematics of Einstein's general theory of relativity. The description of a curved space of four dimensions requires 16 numbers:

$$G_{11}\ G_{12}\ G_{13}\ G_{14}$$
$$G_{21}\ G_{22}\ G_{23}\ G_{24}$$
$$G_{31}\ G_{32}\ G_{33}\ G_{34}$$
$$G_{41}\ G_{42}\ G_{43}\ G_{44}$$

six of which are redundant (because $G_{12}=G_{21}$, $G_{13}=G_{31}$, etc.), leaving ten. These can be arranged in a tetractys-like pattern:

$$\begin{array}{cccc} 11 & 12 & 13 & 14 \\ & 22 & 23 & 24 \\ & & 33 & 34 \\ & & & 44 \end{array}$$

Three separate objects, say a, b, and c, can be arranged in ten distinct triplets:

aaa bbb ccc abc aab abb aac acc bbc bcc

This pattern is found in elementary particle physics, in what is known as the hadron decuplet. Here, u, d, and s are the three elements and stand for three quark flavors. Three quarks make up a hadron—the proton and neutron being the most familiar and ubiquitous. This particular set of ten hadrons falls into this pattern by virtue of the three parameters that are specified: the isospin number along the top, the 'strangeness' (s) number along the left side, and the electric charge (q) diagonally along the right side.

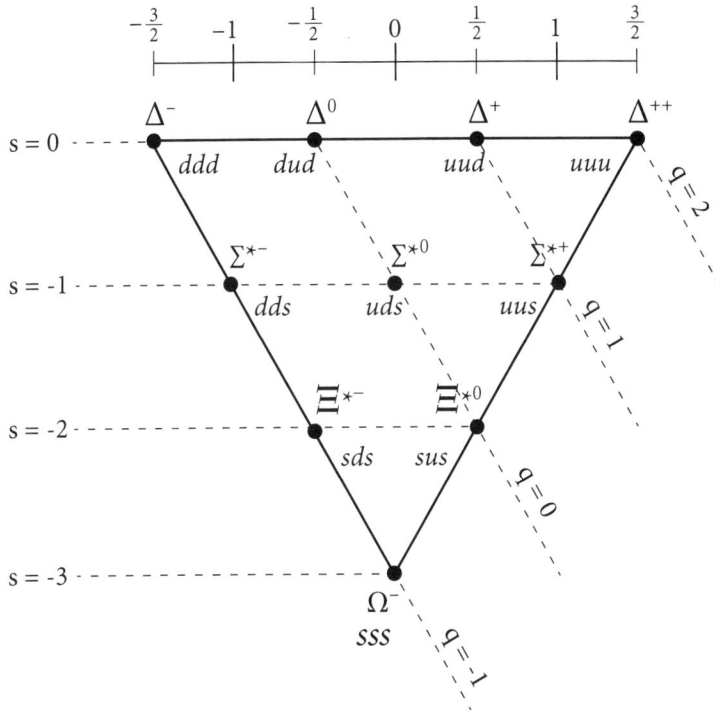

Note the interesting relationship between the numbers 4 (horizontal and diagonal rows), 7 (vertical rows), and 10 (total number of elements).

If one arranges these quarks in what seems like a logical way on an enneagram, the three same-letter triplets on the apices of the triangle, the uds in the center, and the other elements in a logical sequence, one obtains the figure on the following page. The numbers are the electric charge of each hadron and form an interesting pattern as well.

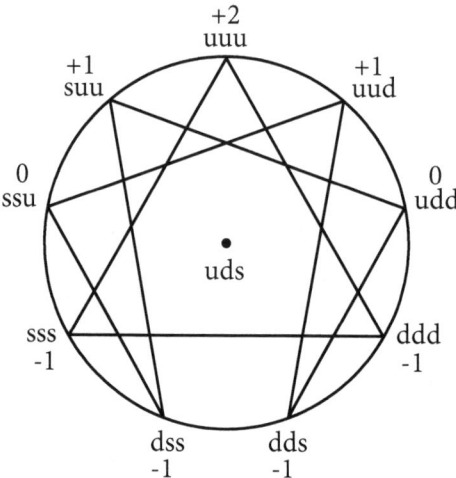

Although no direct connection is evident, the way the pattern segregates into the three single quark triplets at the triangle apices and the triplets composed of one type of quark plus two of another type on the hexagram resembles the pattern of the basic enneagram, with the fractions $\frac{1}{3}$, $\frac{2}{3}$, and $\frac{3}{3}$, repeating decimals of single digits, at the apices of the triangle and the $\frac{1}{7}$ numbers along the hexagram. Whether this is significant in any but a coincidental way is unclear to me.

Six and symmetry

The number 6, being 2 times 3, is important for that reason alone, but it also seems fundamental in the construction of the world. A hexagon inscribed in a circle has sides that correspond exactly to the radius of the circle.

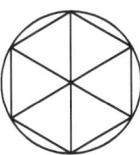

Six-sided figures are found throughout nature, notably in snowflakes, honeycombs, etc. The triangle, square, and hexagon are the only regular polygons that tile the plane without any gaps.

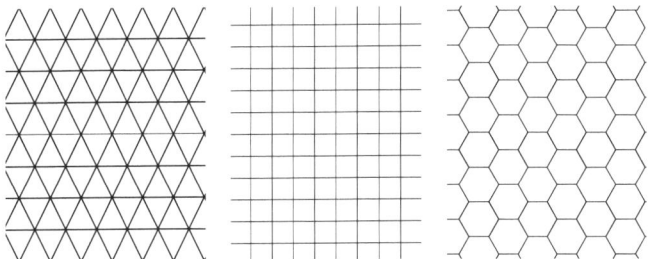

On the other hand, five-sided figures can only tile the plane with irregularities:

Symmetry and its mathematical treatment by group theory (about which more will be said later) have become of paramount importance in modern physics, as the best way to make sense of the structure of the world and its interactions. Technically, a symmetry is a transformation that can be applied to an object that leaves it unchanged in some respect. So, for instance, an equilateral triangle can be rotated 120, 240, or 360 degrees and can be reflected about any of three axes, without altering its appearance.

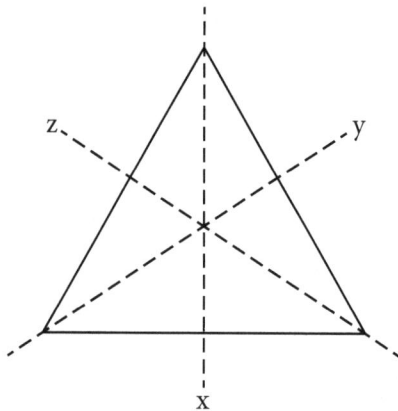

A circle is more symmetric than a triangle because it can be rotated any number of degrees and reflected about any diameter without altering its appearance. If a circle were transformed into a triangle, it would lose this more global symmetry but retain some symmetry, that of the triangle; this is known as broken symmetry. The same phenomenon occurs in many natural circumstances. A pebble thrown onto the smooth surface of a lake breaks the completely symmetric surface of the water, establishing a point that is different from the others, and replaces the smooth surface with a series of concentric circular waves around the point; a greater symmetry has been replaced by a lesser one.

The fact that I do not change into a frog or disappear when I move across the room is a symmetry, involving the stability of the laws of physics despite movement in space and time. Still some things change: my thoughts move along, my view of the room changes, and maybe the temperature of the air against my skin. A perfectly symmetrical world would consist of absolutely empty homogeneous space, and time in which nothing happened, so nothing could exist at all. On the other hand, if there were no symmetry whatsoever, no enduring organized entities could exist either. So the world as we know it must consist of a mixture of symmetry and broken symmetry.

The fact that five-sided figures can tile the plane only with irregularities, whereas 3-, 4-, and 6-sided figures can tile it without any, suggests an aspect of the importance of the number 5: it generates a mixture of symmetry and broken symmetry that leads to an evolving world. Similarly, the golden mean ratio allows growth without change in shape—a mixture of symmetry and broken symmetry. And, as seen on page 27, patterns involving Fibonacci numbers can have self-similarity on different scales without being completely regular.

Furthermore, the pentagon, although it cannot tile the plane with perfect regularity, can be arranged in three dimensions to produce the dodecahedron, one of the five regular (Platonic) solids, with 12 pentagonal faces, 20 vertices, and 30 edges. So, in a sense, five necessitates another dimension for some symmetry to be restored. The evolving world created by broken symmetry involves a proliferation of cosmoses, which is also a proliferation of dimensions. (See below.)

Current speculations in physics postulate that the four fundamental forces of nature—gravity, electromagnetism, and the strong and weak nuclear forces—arose from broken symmetry: at the high temperatures of the early universe that existed just after the 'big bang,' the four forces were one. As water cools and becomes ice, symmetry is broken but other symmetry appears; similarly, as the universe cooled, the fundamental forces precipitated out from one primordial force. This concept depends on advanced mathematical representations of these forces, but the same basic idea is present in the Bible:

> In the beginning God created the heaven and the earth. And the earth was without form, and void; and darkness was upon the face of the deep. And the Spirit of God moved upon the face of the waters. And God said, Let there be light; and there was light. And God saw the light, that it was good: and God divided the light from the darkness. And God called the light Day, and the darkness he called Night. And the evening and the morning were the first day. And God said, Let there be a firmament in the midst of the waters, and let it divide the waters from the waters. And God made the firmament, and divided the waters which were under the firmament from the waters which were above the firmament: and it was so. And God called the firmament Heaven. And the evening and the morning were the second day. And God said, Let the waters under the heaven be gathered together unto one place, and let the dry land appear: and it was so. (Genesis 1:1–9)

A specific version of broken symmetry appears in Gurdjieff's description of the Creation. In order to create our universe, God altered the fundamental laws of nature to make them less symmetric:

> Our Common Father Uni-Being Endlessness, having decided to change the principle of the maintenance of existence of that still unique cosmic concentration and sole place of His Most Glorious Being, first altered the functioning itself of these two primordial sacred laws, and He made the greater change in the law of the sacred Heptaparaparshinokh.
>
> This change in the functioning of the sacred Heptaparaparshinokh consisted in the alteration of what is called the 'subjective action' of three of its 'stopinders'. In one He lengthened the law-conformable duration, in another He shortened it, and in a third, disharmonized it." (BT, pp. 689–690)

Gurdjieff is referring to the second of his two fundamental laws of the universe, the law of seven, also known as the law of octaves, or, as put in *Beelzebub's Tales*, the law of Heptaparaparshinokh. Although the law of three accords with basic logical and scientific principles, the law of seven has not, as such, found its way into the scientific worldview. According to Gurdjieff, the musical major scale is an example, or reflection, of this universal law, and, as already mentioned, there are 'gaps', or irregularities, in this scale that correspond to places in the development of processes where help is needed. (The word 'stopinder' in the above quote refers to the interval between two notes of the scale.)

> The seven-tone scale is the formula of a cosmic law which was worked out by ancient schools and applied to music. At the same time, however, if we study the manifestations of the law of octaves in vibrations of other kinds we shall see that the laws are everywhere the same, and that light, heat, chemical, magnetic, and other vibrations are subject to the same laws as sound vibrations. (ISM, pp. 124–125)

Before elaborating on Gurdjieff's law of seven and the musical scale, I would like to discuss some of the unique properties of the number 7, and its relationships with the number 3. But even prior to that, some attention must be given to some of the progeny of the number 6—12, 24, 60, and 360.

Unlike 10, which is divisible only by 2 and 5, of the numbers less than itself, and 7, which is prime and so has no divisors less than itself (except for 1, which divides everything), 12 and its multiples have an abundance of divisors: 12 is divisible by 2, 3, 4, and 6; 24 has all of these divisors plus 8 and 12; 60 is divisible by 2, 3, 4, 5, 6, 10, 12, 15, 20, and 30; and 360 is divisible by all of the numbers from 1 to 10 except for 7 and additionally by 12, 15, 18, 20, 24, 30, 36, 40, 45, 60, 72, 90, 120, and 180. These features of these numbers have been of great interest to both ancient and modern mathematicians. The 12 signs of the zodiac, the 24-hour day, 60-minute hour, the division of the circle into 360 degrees, each degree into 60 minutes of arc, and each minute into 60 seconds of arc—all originated in the ancient civilizations of Egypt, Sumeria, and Babylonia. The Sumerians and Babylonians used a sexagesimal (base 60) number system for calculations. These numbers have an orderliness which, in a sense, seems the opposite of the irregularities associated with numbers like 5, 7, 10, and the Fibonacci numbers. However, there are relationships: the 12th Fibonacci number is 144 (12 squared), and the Fibonacci numbers fall into groups of 12 and 24 when reduced by theosophical addition, and their last digits repeat after 60 numbers (see p. 55). One again gets a sense of the interactions between order and irregularity, symmetry and broken symmetry.

Six and its progeny generally have this orderly quality, related to their multiple divisors. Six itself was called the marriage number by the Pythagoreans (Michell, p. 52) because it is the product of 2, the first female number, and 3, the first male number. (In general, even numbers were considered female and odd numbers male, for a variety of reasons.) It is also the first so-called 'perfect' number, being equal to the sum of its divisors (1+2+3).

Twenty-four is one of John Baez's favorite numbers (Baez). In his lecture, he points out a number of intriguing properties of the numbers 12 and 24. A square pyramid with a base of 24 elements on a side has 24^2 elements in the base, followed by 23^2 on the next level, 22^2 on the next, etc., until at the top there is a single element, for a total of $24^2 + 22^2 + 21^2 + \ldots\ldots + 1$ elements, which is equal to 70^2 elements, a curious connection between 24, 7, and 10. Furthermore, this is the only square pyramid with a total number of elements that is a square number. Baez also somewhat mysteriously states that 24 is important in the scheme of things simply because it is 6 times 4. Among other mutual properties, 6 and 4 are the numbers of sides of the regular polygons that, along with the triangle, tile the plane without any gaps (see p. 31).

In the enneagram, 6 is represented by the hexagram. The hexagram also relates 2 and 3: it connects the two sides of the enneagram, right and left, and it straddles the 3 and 6 points, enabling the passage through the intervals in the scale (to be discussed later).

7

The unique properties of the number 7 and its relationships with the number 3

The number 7 seems to be a favorite of humanity, appearing in numerous places, from the days of the week and the colors of the rainbow to the seven seas, the seven wonders of the ancient world, the seven samurai, and the seven dwarfs. It also has a large number of unique mathematical properties, long lists of which can be found in Wikipedia and other internet sites. Many of these properties are related to each other and to those which will be discussed below; others seem somewhat arbitrary, although hidden relationships may exist there also.

Philo Judaeus, in *On the Creation of the World*, said:

> And such great sanctity is there in the number seven, that it has a preeminent rank beyond all the other numbers in the first decade. For the other numbers, some produce without being produced, others are produced but have no productive power themselves; others again both produce and are produced. But the number seven alone is contemplated in no part. And this proposition we must confirm by demonstration. Now the number one produces all the other numbers in order, being itself produced by absolutely no other; and the number eight is produced by twice four, but itself produces no other number in the decade. Again, four has the rank of both, that is, of parents and offspring, for it produces eight when doubled, and it is produced by twice two. But seven alone, as I said before, neither produces nor is produced, on which account the philosophers liken this number to Victory, who has no mother, and to the virgin goddess, whom the fable asserts to have strung from the head of Jupiter: and the Pythagoreans compare it to the Ruler of all things.

This is one, ancient, explanation of the uniqueness of the number 7.

Another of course is the unique pattern, among the first 10 numbers, formed by the decimal expansion of $1/7$, 0.142857... (chapter 2). In addition to the features already discussed above, one can find in Wikipedia several interesting mathematical relationships of 142857:

$$0.142857\ldots = 0.14 + 0.0028 + 0.000056 + 0.00000112\ldots$$
$$= {}^{14}\!/_{100} + {}^{28}\!/_{100^2} + {}^{56}\!/_{100^3} + {}^{112}\!/_{100^4} + \ldots + {}^{7\times 2^n}\!/_{100^n} + \ldots$$
$$= {}^{7}\!/_{50} + {}^{7}\!/_{50^2} + {}^{7}\!/_{50^3} + \ldots + {}^{7}\!/_{50^n} + \ldots,$$

which relates 100, 50, and 7.

$$\tfrac{1}{7} = 0.1 + 0.03 + 0.009 + 0.0027 + 0.00081\ldots$$
$$= {}^{3^0}\!/_{10^1} + {}^{3^1}\!/_{10^2} + {}^{3^2}\!/_{10^3} + \ldots + {}^{3^{n-1}}\!/_{10^n} + \ldots \text{ This relates 3 and 7.}$$

Perhaps the simplest appearance of 7 comes about owing to the seemingly magical feature of the hexagon mentioned above: a regular hexagon fits perfectly into a circle, and the sides of the hexagon are equal to the radius of the circle. For the same reason, six identical circles fit perfectly around a seventh.

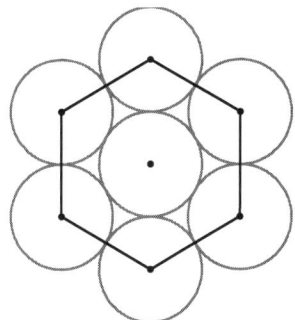

More circles can be added, resulting in the heptagonal numbers, with formula

$$n^3 - (n-1)^3$$

These patterns have 1, 7, 19, 37, 61, etc., circles:

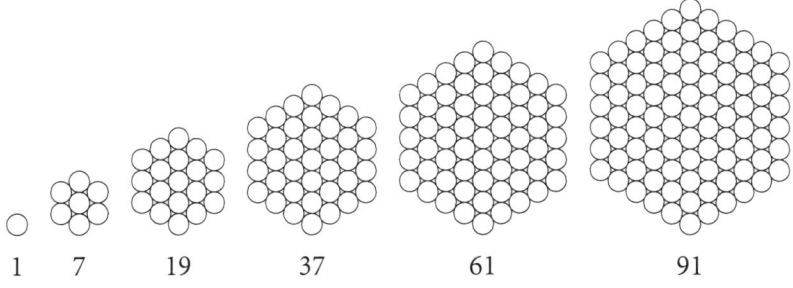

The pattern with 61 circles is describable as six tetractys around a central circle:

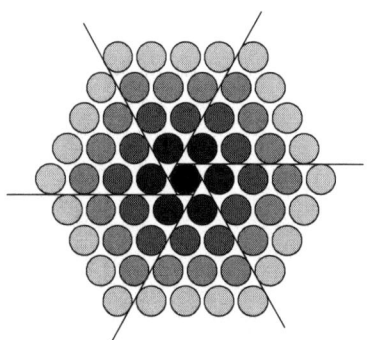

In three-dimensional space, if one counts the center as one direction, there are seven directions: center, up, down, right, left, forwards, backwards.

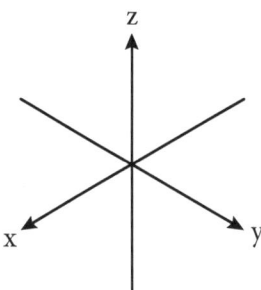

A cube, with six faces, viewed by looking straight at one of its vertices and flattened, looks like a hexagon:

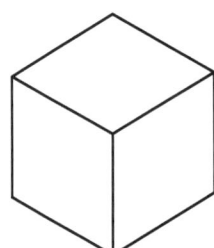

In a high school math text, one can find the following:

Three random straight cuts of a circle will produce at most seven regions, and three circles intersecting each other also results in seven regions. These particularities also relate three and seven.

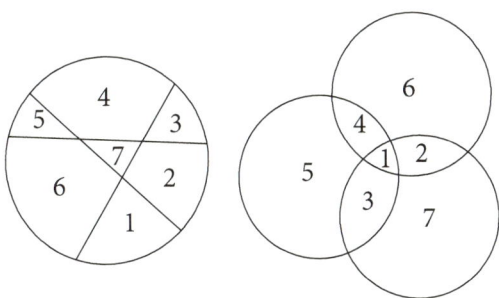

The tetractys can be bisected horizontally into three and seven points. (Beke, p. 36)

The Greek lambda, about which more will be discussed in chapter 19, has two triads, one of powers of 2 and one of powers of 3, along with the 1 at the top, for a total of seven numbers. (Beke, p.46)

```
              1
          2       3
        4           9
      8              27
```

The number 7 has many other interesting properties that may contribute to its importance. As pointed out by Arthur M. Young (Young, pp. 259–282), seven elements can be arranged in seven triplets such that each triplet shares one and only one element with each of the other triplets.

```
    a b c d e f g
        abd
        bce
        cdf
        deg
        efa
        fgb
        gac
```

This can be depicted as seven triangles connecting seven points around a circle, and as a result, every point on the circle is connected to all the others:

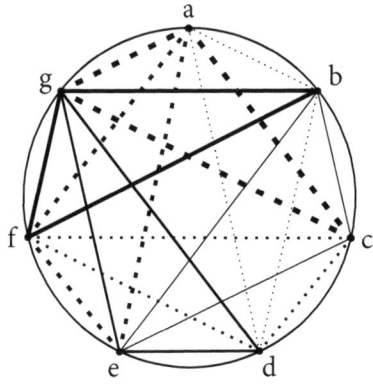

If one has only six elements, only four such triads are possible. If there are more than seven elements, it is not possible to make such triads.

Number of elements: 1 2 3 4 5 6 7 8
Number of triads: 0 0 1 1 2 4 7 0

These relationships also appear in what is called the Fano plane, about which more will be discussed later. Each line (the circular line in the center is considered a line) connects three points; each point belongs to three lines; and each line shares one and only one point with each of the other lines.

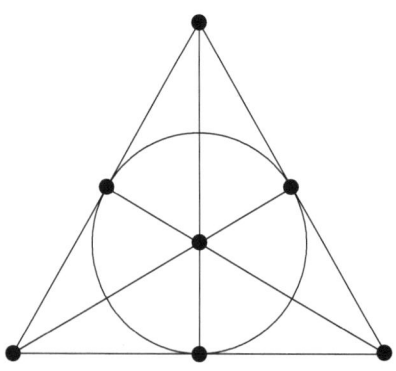

Triples of 0s and 1s can be arranged in the Fano plane, such that along any line, two of the triples add up to the third, modulo 2 (0+0=0, 1+0=1, 1+1=2=0).

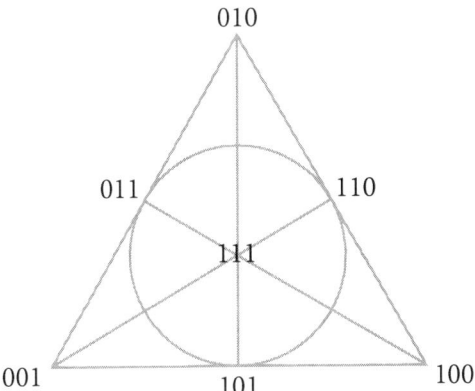

The enneagram depicts interesting relationships between 2, 3, 6, 7, and 9. It is bilaterally symmetrical. Three points form the triangle. Six points form the inner figure. With the addition of the point 9, the seven-note scale is formed:

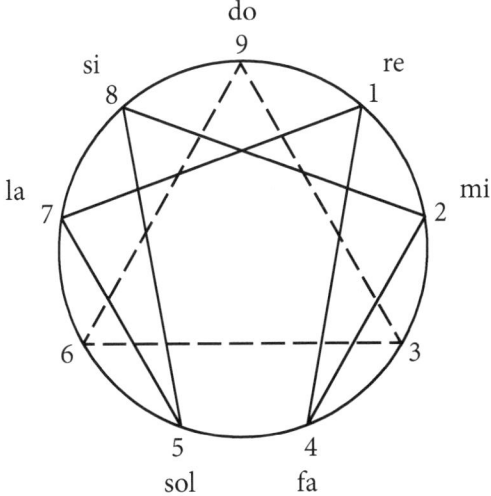

Arranging the triads of 0s and 1s from the Fano plane along the seven notes of the scale, one could obtain:

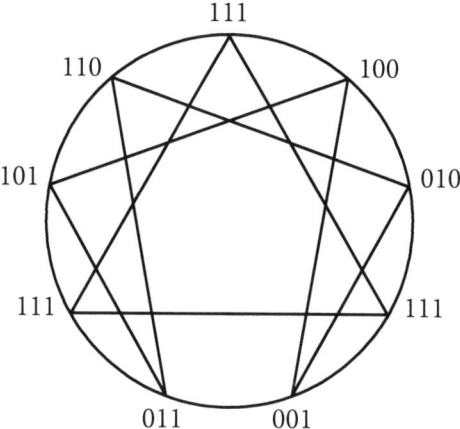

Here, 111 is at the points of the triangle, and the right side contains triples with two 0s and the left side triples with two 1s.

These triples of 0s and 1s also correspond to the eight Chinese trigrams of broken and unbroken lines (yin and yang), by making 0 correspond to a broken line and 1 to an unbroken one. There are eight trigrams, rather than seven, because 000 is now included:

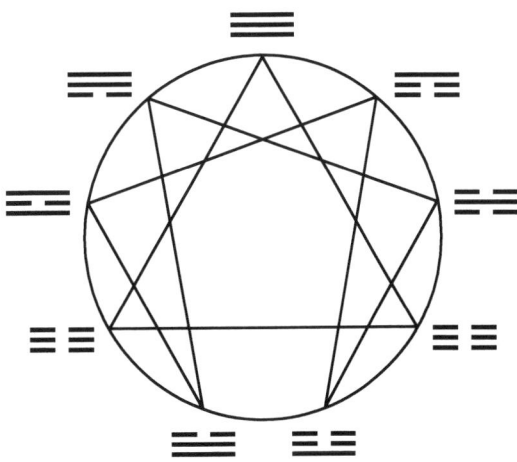

Young also points out that on the surface of a torus, seven colors are required to color a map so that no two 'countries' sharing a border have the same color. On a sphere, only four colors are needed. Equivalently, on a torus, one can connect seven points each to all the others without the connecting lines crossing each other. The torus, and the portion of it called a saddle, are of great importance in mathematics and physics. The enneagram can also be seen as a torus: Gurdjieff stated that each note on the circle contains an entire 'inner octave' on another scale, reflecting a hierarchy of cosmoses (ISM, p. 289).

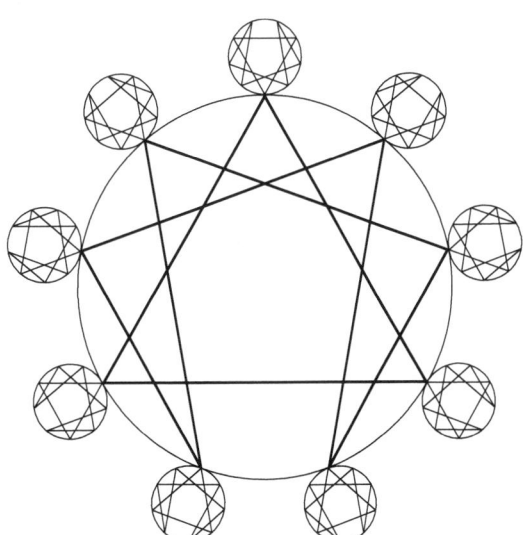

If one regards the inner octave enneagrams to be perpendicular to the larger one, which makes sense since the inner octaves can be regarded as representing another level of cosmoses, and cosmoses are related to each other as orthogonal dimensions ("The Teaching of the Cosmoses." In MiC, pp. 101–114; and also later within this text), one obtains a toric figure.

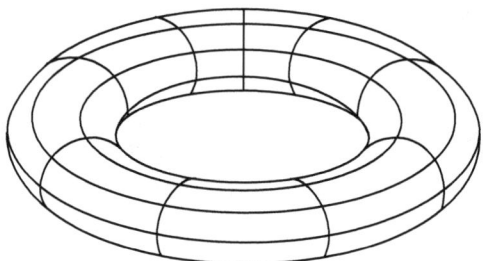

The law of seven and the seven-tone scale

It is well known that a doubling of the frequency of sound vibrations results in a note that is perceptually the same as the original note, only higher. While this sounds paradoxical, it is evident and universal in human experience. For instance, the note la (A in American notation), which consists of 220 cycles per second (Hz) of air pressure vibrations, sounds like the same note as the note la which is 440 Hz—only the latter has a higher pitch. So in the case of octaves (the second la is an 'octave' or eight notes above the first in a seven-tone scale system—see below), there is a doubling of the vibration rate, a 2:1 ratio. It is also evident to the human ear that other small integer ratios of vibrations result in harmonious sounds, although not the "same" sound, as is the case with octaves. Thus, ratios of 3:2, 4:3, and 5:4 produce harmonious sounds, whereas ratios of larger adjacent numbers, like 9:8 or 10:9, sound progressively more dissonant. The ratio 6:5 is at the edge of sounding dissonant. The discovery of the relationship of small number ratios to musical harmoniousness is attributed to Pythagoras, although it is likely that it was known in even more ancient times. The seven tone 'major' (in Western musical terminology—also called 'ionian') musical scale incorporates these harmonies and is based on a succession of ratios. There are several slightly different ways to create the major scale, about which more will be said later, but the one described below is called "just intonation" and was thought by Gurdjieff to represent the universal law of seven.

Do represents the starting pitch, and the notes re, mi, fa, sol, la, and si make up the rest of the seven-tone scale. The notes are also known to musicians by their position in the scale, so for instance, sol is a 'fifth' above do, being the fifth note of the major scale. The eighth note, double the first in vibrations, is the octave, which sounds like the first note, and so restarts the cycle at a higher level. If we take do as 1, the subsequent ratios of vibration rate are

1	$9/8$	$5/4$	$4/3$	$3/2$	$5/3$	$15/8$	2
do	re	mi	fa	sol	la	si	do

Played together, do and mi, do and fa, do and sol, and do and la sound very harmonious, although each harmony has a different quality, while do and re, and do and si do not.

These fractions represent the vibration rate of each note compared to the first. The difference between one note and the next is found by dividing the ratios. So, $5/4$ divided by $9/8$ is $40/36$ or $10/9$. Thus the ratios of the vibrations of each note relative to the one below are as follows:

$$\text{Between re and do: } 9/8 : 1 = 9/8$$
$$\text{Between mi and re: } 5/4 : 9/8 = 10/9$$
$$\text{Between fa and mi: } 4/3 : 5/4 = 16/15$$
$$\text{Between sol and fa: } 3/2 : 4/3 = 9/8$$
$$\text{Between la and sol: } 5/3 : 3/2 = 10/9$$
$$\text{Between si and la: } 15/8 : 5/3 = 9/8$$
$$\text{Between do and si: } 2/1 : 15/8 = 16/15$$

Between most notes and the previous one, there is an increase in vibration rate of either $9/8$ or $10/9$, but between fa and mi and between do and si, there is only an increase of $16/15$. As a result, in Western music, another note is placed between each two notes separated by $9/8$ or $10/9$, called either a sharp of the lower note or a flat of the upper one, but between mi and fa and si and do, there is no extra note. The resulting scale (now called chromatic) then has 12 notes:

do	do#	re	re#	mi	fa	fa#	sol	sol#	la	la#	si
	(do sharp)										

At the places in the initial seven-tone scale where there is only a $16/15$ increase, there is a slowing of the increase in vibration rate, which Gurdjieff regarded as places (the 'gaps') where a developing process—of any sort, because the musical scale represents a universal law—will deviate from its original course, unless an additional impulse is provided at those points to maintain the original momentum.

> This law shows why straight lines never occur in our activities, why, having begun to do one thing, we in fact constantly do something entirely different, often the opposite of the first, although we do not notice this and continue to think that we are doing the same thing that we began to do.
>
> Such a course of things, that is, a change of direction, we can observe in everything. After a certain period of energetic activity or strong emotion

or a right understanding a reaction comes, work becomes tedious and tiring; moments of fatigue and indifference enter into feeling; instead of right thinking a search for compromises begins; suppression, evasion of difficult problems. But the line continues to develop though now not in the same direction as at the beginning. (ISM, pp. 128–129)

For processes to develop properly, an additional input, or 'shock,' is required at the intervals where there is a slowing of the increase in vibration rate. Gurdjieff used the metabolism of food as an example. The breakdown of food macromolecules (essentially starches, proteins, and fats) into smaller molecules, with the concomitant release of chemical energy for the body's use, begins in the mouth, stomach, small intestine, liver, and other organs but can only proceed so far biochemically until oxygen, carried in the blood, is added to the mix. Then the breakdown of food molecules can proceed all the way to carbon dioxide and water, with a much greater total release of energy. The first part of the process is called anaerobic (without air) metabolism, and the second, aerobic metabolism. Gurdjieff regarded anaerobic metabolism as taking the do of undigested food up to the level of mi, at which point the entry of air or oxygen gives the necessary boost for the process to go further, to si. For the transformation of energies to proceed further, from si to the next do, another 'shock' is needed.

For Gurdjieff, the change in the law of seven, from a completely symmetrical self-contained scale with equal intervals to one with altered intervals, resulted, through broken symmetry, in the creation of the now existing universe. Importantly, it also made processes in the universe interdependent, in that any given process requires 'outside help' to develop properly.

Just as with the pebble in the lake (see page 32), the partial breaking of symmetry results in a cornucopia of new phenomena and leads to our world, perched precariously between the two extremes, of perfectly symmetrical absolutely empty and eventless spacetime, and completely asymmetric and irregular total chaos.

* * *

Why does a musical octave sound so harmonious that the two notes sound essentially like the same note? One factor may be related to the overtone patterns of musical sounds. A vibrating string can, and does, vibrate as a whole, in halves, thirds, fourths etc., and in any subdivision of the string that allows the two fixed ends to stay relatively still:

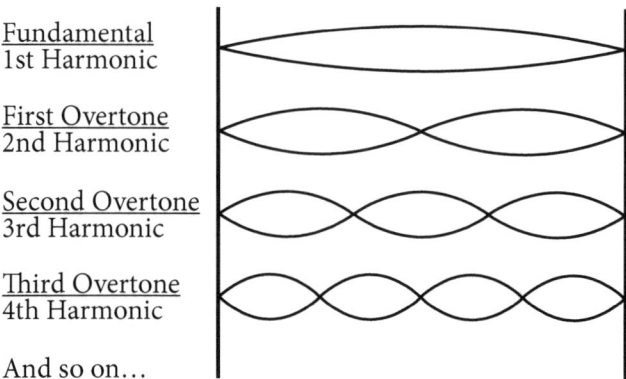

Fundamental
1st Harmonic

First Overtone
2nd Harmonic

Second Overtone
3rd Harmonic

Third Overtone
4th Harmonic

And so on…

The corresponding notes will be the fundamental, corresponding to the string vibrating as a whole, and then the octave, because each half of the string is vibrating twice as fast. The second overtone is a note with three times the vibration rate of the fundamental, which then has a 3:2 relationship with the octave. This note corresponds to the fifth above the octave. The third overtone, with four times the vibration rate of the fundamental, is an octave above the octave, or two octaves above the fundamental. The fourth overtone, with five times the vibration rate of the fundamental, has a 5:4 relationship with the second octave, and thus corresponds to the note mi. And so on. All these notes are present to varying degrees in the sound of a vibrating string; the fundamental is generally the loudest, and the overtones generally become less prominent as the number of subdivisions of the string increases. The specific proportions of the various overtones—determined by the characteristics of the string and the instrument of which it is a part—produce the particular 'timbre,' or quality, of the sound of that string on that instrument.

The same phenomena apply, slightly differently, to the vibrating air columns in wind instruments.

If two notes are sounded together, their overtones may coincide to varying degrees. If, for example, one note has a fundamental with a vibration rate of 100 Hz, its overtones will be 200 Hz, 300 Hz, 400 Hz, 500 Hz, 600 Hz, etc. If the second note is an octave above the first, its fundamental will be 200 Hz, and its overtones 400 Hz, 600 Hz, etc. Every other overtone of the first note is the same as an overtone of the second note, so it would make sense that this contributes to the harmoniousness of the sound.

In the case of two notes a fifth apart, do and sol, there is also a coincidence of overtones, but not as much as with the octave. If the first note's fundamental is 100 Hz and the second's is 150 Hz, the second note's overtones will be 300 Hz,

600 Hz, etc., and the overtones of the first note will coincide with those of the second every third overtone. In the case of the fourth, do and fa, there will be coincidence of overtones every fourth overtone of the lower note, and in the case of the third, do and mi, every fifth.

FUNDAMENTAL	OCTAVE	FIFTH	FOURTH	THIRD
100 Hz		150	133 ⅓	125
200	200*		266 ⅔	250
300		300*		
400	400*		400*	375
500		450	533 ⅓	500*
600	600*	600*		625
700		750	666 ⅔	750
800	800*		800*	
900		900*		875
1000	1000*			1000*

The asterisks mark the overtones that coincide with the fundamental's overtones.

Note that in the case of the octave, *all* of the higher note's overtones coincide with overtones of the lower note, perhaps explaining why they sound like the "same" note. The harmoniousness of the other intervals is felt by many listeners to follow the same decreasing order as the amount of coincidence of their overtones: fifth, fourth, third.

While all this is of great interest with regard to musical harmonies, does it have broader applicability, as suggested by Gurdjieff's idea that the musical scale represents a fundamental law of the universe?

The electrons in atoms are also described mathematically as vibrations, although these are more complex than the vibrations of a string, in part because they are three dimensional and also because they are not vibrations of anything that we regard as material, but rather vibrations of 'probability amplitudes.'

Vibrational modes of hydrogen (electron probability amplitudes.)

One can regard the relative affinities of atoms, their tendency to combine or not, as analogous to the relative harmoniousness of musical tones. The waveforms that result when multiple tones are combined are more complex than those of the original tones but remain in a repetitive coherent pattern for as long as the combination lasts. Similarly, a molecule made of a combination of atoms forms a more complex but stable, for a time, vibratory pattern.

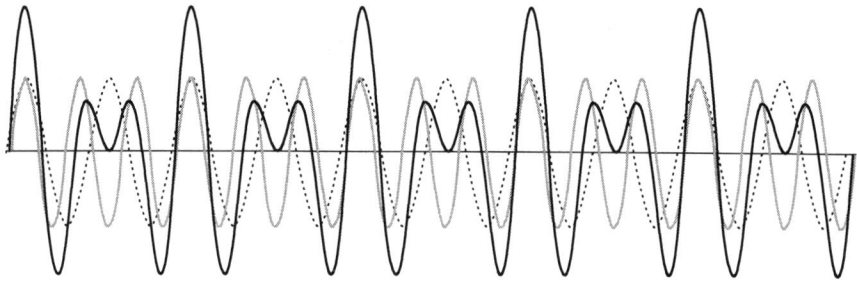

The dotted and grey sine waves are in a 3:2 relationship. The black wave is the waveform that results when the two are "played" together. As more component sine waves are added, more complex vibratory patterns are formed.

It is also evident that the different energy levels, or excited states, of atoms and molecules correspond precisely to overtones, although in a greater dimensionality than the overtones of a vibrating string. The reason there are 'quantum leaps' from one state to another, without smooth gradual transitions, is that the vibrations corresponding to a given state have to 'fit' into a corresponding orbital, just as the overtones of a string have to obey the constraint that the two ends of the string be relatively still points.

* * *

Gurdjieff taught that human beings are 'three-brained,' consisting of a body-centered moving/instinctive brain, an emotional brain, and an intellectual brain, as opposed to other mammals, which he regarded as lacking the third, intellectual, brain, and creatures such as worms, which have a moving/instinctive brain only ("Does Man Have Three Brains." In MiC, pp. 147–186). In parallel, there are three kinds of food, corresponding to our three brains: ordinary food, air, and impressions, of progressively finer materiality.

Everything in the world is material and—in accordance with universal law—everything is in motion and is constantly being transformed. The direction of this transformation is from the finest matter to the coarsest, and vice versa.

Between these two limits there are many degrees of density of matter. (VRW, p. 209)

To each level of materiality, Gurdjieff assigned a number: ordinary food was matter of density 768; water, 384; air, 192; and impressions, 48. As described above, air is needed to allow the breakdown of ordinary food to get past mi. In addition, impressions are needed to help the development of the air octave get past its own mi. These ideas were put on the enneagram by Gurdjieff and Ouspensky (ISM, p. 377):

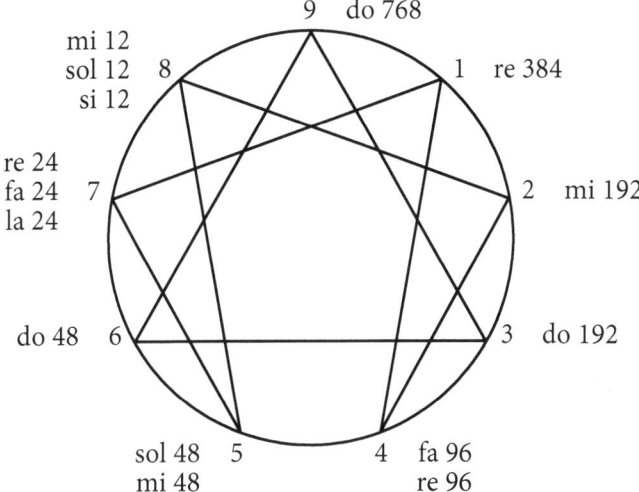

Air (do 192) enters where there is a gap between mi and fa in the development of the food octave, corresponding to where there is a slowing of the increase of vibration rate; thus air provides the help needed for the food octave to get past the gap and develop to the level of si 12. At the same time, the air octave, beginning at do 192, can develop to mi 48. Here help is needed for this octave to develop further, and impressions (do 48) enter in the interval between mi and fa of the air octave. Gurdjieff regarded the impressions needed at point 6 of the enneagram as not merely the usual impressions that allow us to function ordinarily but as impressions accompanied by self-awareness, or 'self-remembering,' about which more will be discussed later. The metabolism of the three kinds of food, aided by the two 'shocks,' reaches si, sol, and mi of materiality 12, which Gurdjieff states corresponds to the materiality of atomic hydrogen (ISM, p. 175). This can be interpreted as the level of electromagnetism, as an atom of hydrogen consists of a single proton and a single electron, and hydrogen dissociates in the body into free positively charged protons and negatively charged electrons, which provide the body (via complex electromagnetic processes) with energy. Gurdjieff also associates si 12 with sexual energy.

The metabolism of food, then, can be regarded as a process whereby a succession of temporarily stable complex vibratory patterns occurs, leading to finer and finer materialities and energies. An organism is itself an extremely complex coherent but partially and continually changing vibratory pattern. When a critical amount of coherence is lost, the organism dies.

* * *

A discrepancy becomes evident between the placement of the 'gaps' according to the just-intonation scale and their placement in the enneagram: in the major scale, the two smaller intervals are between mi and fa and si and do, whereas in the enneagram, the second smaller interval is between sol and la. Gurdjieff pointed this out himself in his discussion of the enneagram and attached a significance to this discrepancy which was not clearly elucidated (at least the elucidation [ISM, pp. 291–293] is not clear to me).

> The apparent placing of the interval in *its wrong place* itself shows to those who are able to read the symbol what kind of 'shock' is required for the passage of si to do. (ISM, p. 291)

There is, however, an interesting connection with the notes of the scale. Many of the notes of the major "just" scale are present in the overtone series. Sol is the second overtone and mi the fourth. The first fifteen overtones, for a fundamental of 100 Hz, are as follows:

100	do
200	do
300	sol ($3/2$ of second do)
400	do
500	mi ($5/4$ of third do)
600	sol ($3/2$ of third do)

700	not a note of the just scale
800	do
900	re ($9/8$ of fourth do)
1000	mi ($5/4$ of fourth do)
1100	not in the just scale
1200	sol ($3/2$ of fourth do)
1300	not in the just scale
1400	not in the just scale
1500	si ($15/8$ of fourth do)
1600	do

The coincidence of many of the notes of the overtone series with the notes of the just scale gives support to the idea that the just scale arises "naturally" from the laws of vibrations, and that its significance may extend beyond music. But the notes fa and la are not in the overtone series. This is because the ratios representing them have a 3 in the denominator, whereas the notes in the overtone series all have a multiple of 2 in the denominator, necessitated by the fact that the octaves consist of doublings of the vibration rate. So it is interesting that in the enneagram the notes fa and la occur right after the intervals where a 'shock' is needed.

The fact that the two notes following the shocks have a 3 in the denominator rather than a 2, and consequently are not in the overtone series, may relate to why the shock is needed in these places. Comparing the two versions of the placement of the shocks, they agree on the placement of the first shock—it is both before a smaller increase in vibration rate according to the just scale and before a note with a 3 in the denominator according to the enneagram—but they disagree on the placement of the second shock. If this were understood, it might well provide a clue as to the nature of the second shock, as Gurdjieff stated, and also relate to the difference between the two shocks, called by Gurdjieff the 'mechano-coinciding mdnel-in' and the 'intentionally actualized mdnel-in.' (See later in the text.)

Another numerical aspect of the "wrong" placement of the second gap in the enneagram is discussed on page 60.

In the third part of the enneagram that depicts the three kinds of food and their interactions (points 7 and 8), three note chords are produced: re-fa-la and mi-sol-si. Three note chords are fundamental in music and represent another level compared to two note harmonies, which themselves are a level above single notes. So this enneagram also represents the progressive development of the possibilities of three-brained beings. A more complete resonance with the structure of the universe becomes possible when the energies from the three kinds of food become harmonized. This is another way of saying that the three brains have to be simultaneously and harmoniously active for true perception to be possible.

* * *

In the enneagram, there is a separation between the multiples of 3 and the other digits, resulting from the absence of multiples of 3 in the fraction $1/7$. This segregation of multiples of 3 from the other digits appears in many places if one uses theosophical addition in the decimal number system (modular arithmetic modulo 9). For instance, powers of 2:

| 2 | 4 | 8 | 16 | 32 | 64 | 128 | 256 | 512 | 1024 |

have the theosophical sums:

| 2 | 4 | 8 | 7 | 5 | 1 | 2 | 4 | 8 | 7 |

and so on, endlessly repeating 248751, the same numbers as in the decimal expansion of $1/7$, although in a different order.

Powers of 3, however, give only 9s, after the first, 3^1:

3	9	27	81	243	729	2187	6561	19683
3	9	9	9	9	9	9	9	9

In other series, although all the digits are included, the multiples of 3 stand out from the others, appearing at regular intervals:

Multiples of 2:	2	4	6	8	10	12	14	16	18	20
Modulo 9:	2	4	6*	8	1	3*	5	7	9*	2

Multiples of 3 themselves only give 3s, 6s, and 9s:

3	6	9	12	15	18	21	24	27	3	33
3	6	9	3	6	9	3	6	9	3	6

Even the Fibonacci series behaves in this way under theosophical addition, segregating 3, 6, and 9 from the other digits:

Fibonacci #:	Theosophical sum:	Fibonacci #:	Theosophical sum:
1	= 1	987	= 6
1	= 1	1597	= 4
2	= 2	2584	= 1
3	= 3	4181	= 5
5	= 5	6765	= 6
8	= 8	10946	= 2
13	= 4	17711	= 8
21	= 3	28657	= 1
34	= 7	46368	= 9
55	= 1	75025	= 1
89	= 8	121393	= 1
144	= 9	196418	= 2
233	= 8	317811	= 3
377	= 8	514229	= 5
610	= 7	832040	= 8

etc.

The reduced sequence thus formed can be arranged in two rows of twelve numbers, after which it repeats:

112358437189
887641562819
112358437189
887641562819

This series has several interesting properties. First is the fact that after 24 numbers it repeats itself. Second is the arrangement of the digits: the numbers 3, 6, and 9 occur in a regular pattern, every fourth digit. The other digits—1, 2, 4, 5, 7, 8—are interspersed in between. Third, the two rows of twelve numbers form complements: each vertical pair adds up to 9, except for the last two 9s, which, by adding up to 18, reduce to 9 also.

Related to this repeating pattern is another: the last digit of a Fibonacci number repeats after sixty numbers. So, for instance, the second Fibonacci number is 1 and the 62nd is 4052739537881, also ending in 1; the 63rd Fibonacci number ends in 2, etc.

Of interest with regard to these patterns is that the Fibonacci series, intimately connected with fiveness by way of its association with Ø, generates repeating patterns involving 12, 24, and 12×5=60.

We have seen that powers of 2 and powers of 3 never coincide. This also relates to the musical scale and to the cause of the creation of the modern 'tempered' scale.

Musicians are familiar with the 'cycle of fifths,' one aspect of which is that twelve successive fifths correspond to seven octaves. Sol is the fifth (five notes up the scale) of do, and re the fifth of sol. The sequence—always maintaining a 3:2 ratio between successive notes, thus necessitating the use of the chromatic 12-tone scale—is

do   sol   re   la   mi   si   fa#   do#   sol#   re#   la#   fa   do

However, the relation between twelve fifths and seven octaves is not perfect, because powers of 3 and powers of 2 never coincide exactly. $3/2$ multiplied by itself 12 times—$(3/2)^{12}$—equals $531441/4096$, or 129.746, whereas seven octaves equal $(2)^7$ or 128. This difference—$129.746/128$, or $531441/524288$—is known as the Pythagorean comma. This small misalignment is related to the fact that mistuned notes occur if one changes from one key to another in an instrument tuned according to the just system. If one tunes an instrument to the just scale for, say, do major, putting the usual sharp semitone between each of the larger intervals but not between mi and fa or si and do, the resulting scale will sound perfect. But if one tries to modulate to another major scale, say sol major, using the same notes, difficulties appear. Suppose that do is 100 Hz, then la, the sixth tone in the do major scale, is $100 \times 5/3 = 166.66$. If one switches to sol major, la is now the second tone in that scale, and should be $150 \times 9/8 = 168.75$, but the only available note is 166.66. So there is a slight mistuning, and this can be worse for other notes and other modulations. As a result, the tempered scale was created, with 12 logarithmically equal semitones. The resulting scale is slightly off compared to the just scale for all harmonies except the octave, but the discrepancies are minimal, spread across all the intervals, and the same regardless of modulating to another fundamental. Bach's well-tempered clavier compositions were a celebration of this new capacity for modulation.

The slight difference between the twelfth fifth and the seventh octave, if they are played together, produces a slight dissonance, a tension, and creates another vibration, whose frequency is that of the coinciding peaks of the two waves.

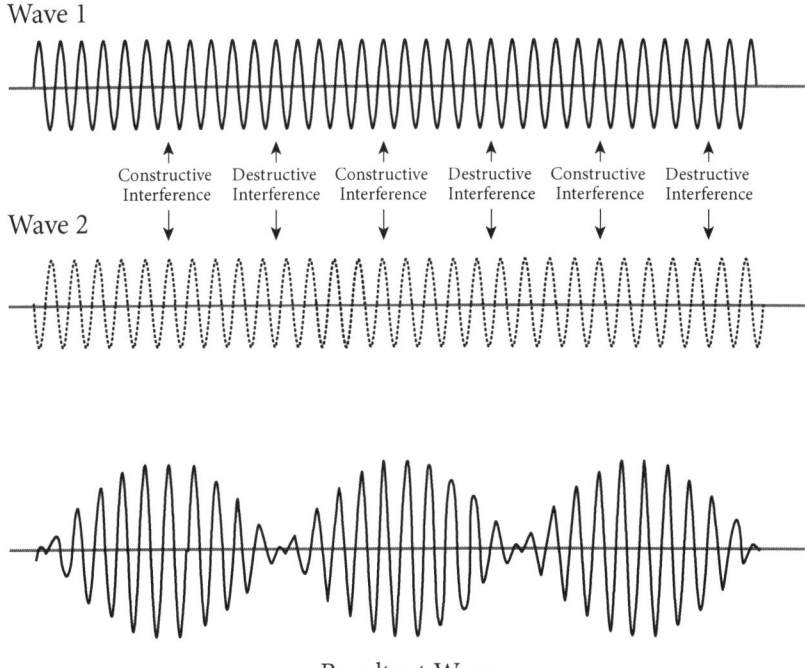

Resultant Wave

In this example, one wave has a frequency of 10 cycles per second, and the other, of 11 cycles per second. The two waves played together will create a third wave whose amplitude at each moment is the sum of the amplitudes of the two component waves. When the two waves are simultaneously at their peak, the amplitudes add; when one is at its peak and the other at its trough, the waves cancel each other. This creates what are called beats, another frequency corresponding to the coinciding peaks of the two waves. In this example the beat frequency is one cycle per second, the same as the difference between the two tones.

In the case of the difference between the 12th fifth and the seventh octave, the beat frequency is even slower than in the example above, and much slower compared to the frequency relationships of the main harmonies, e.g. $^{129.75}/_{128}$ as opposed to $^{3}/_{2}$ or $^{5}/_{4}$. In each case the denominator indicates how many vibrations of the lower note occur before there is an exact coincidence of the peaks of the two waves. This dissonance can be regarded as a tension or force between the two notes, which creates a vibration on another level, that is to say, on a different temporal scale. Although there is no obvious direct connection, this seems analogous to the residual forces, resulting from imperfect symmetry, that create the different levels of interaction in the world:

> The electromagnetic force binds electrons and nuclei to make atoms. The atoms, although they are electrically neutral, interact through a residual electromagnetic force to form molecules. The strong force binds quarks to make protons, neutrons, and all other hadrons, and the residual strong force between protons and neutrons is the so-called nuclear force that binds them into nuclei. (Haber)

These speculations suggest that the asymmetries that led to the creation of the now existing world derive, at least in part, from asymmetries inherent in the natural number system. This idea is central to the worldview of the Pythagoreans—"all things are number"—and of Gurdjieff.

* * *

Although the connections are not entirely clear, it is tempting to relate the discrepancy involving the relationship of twelve fifths (sol) to seven octaves (do) to Gurdjieff's special treatment of the note sol as a similar unfolding of phenomena on different levels is suggested.

That slight discrepancies, or asymmetries, lead to phenomena on another level is seen in a number of instances. Slightly mistuned notes result in "beats," vibrations of much lower frequency, as described above. Slight differences in the views from the two eyes, because of the distance between them, are transformed by the brain into a direct perception of depth, another dimension, that is not directly perceivable by one eye alone ("Shadows of the Real World." In MiC, pp. 65–75). And, as suggested by the above quote, slight imbalances of forces on one level result in forces on another. There is an intimate relationship between changes of scale (of size or frequency), dimensions, and the idea of cosmoses. As we go from the stronger nuclear forces to the weaker atomic and then even weaker molecular ones, there is a progressive change of level, and these differences correspond to a significant degree to the cosmoses as described above: in the sun, nuclear forces prevail, while in and on the earth, atomic forces become more relevant. Molecular forces are prevalent in organic life.

Gurdjieff stated that cosmoses are related to each other as 'zero to infinity,' thus directly comparing them to dimensions (ISM, p. 206). As discussed in more detail elsewhere ("The Teaching of the Cosmoses." In MiC, pp. 101–114), in relation to cosmoses, it would be better to say '*almost* zero to infinity,' in contradistinction to abstract mathematical dimensions.

> Looked at from our scale, the cell is *almost* a point, the earth is *almost* flat, and the speed of light is *almost* infinite. We could postulate that the real dimensions of the world are *almost* perpendicular. (MiC, p. 111)

Gurdjieff ascribes a pivotal role to the note sol in relation to the creation of the cosmoses, although his treatment of the subject is somewhat difficult to decipher. He states that the changes made by the Creator in the two intervals—between mi and fa, and between si and do—led to a disharmonization of a third interval, between sol and la:

> As regards the third stopinder which was changed in its subjective action and which is fifth in the series and called 'harnel-aoot,' its disharmony came about by itself, simply as a result of the change of the other two stopinders.
>
> This disharmony of its subjective functioning, resulting from its 'asymmetry' in relation to the whole process of the sacred Heptaparaparshinokh, consists in the following:
>
> If the completing process of this sacred law flows in conditions in which it is subject to many 'extraneously caused' vibrations, its functioning produces only external results.
>
> But if this same process takes place in absolute quiet, in the absence of any extraneously caused vibrations whatever, all the results of the functioning of this stopinder remain within that concentration in which the process is completed, and these results only become perceptible to the outside on direct and immediate contact with it.
>
> But if during the functioning of this process neither of these two sharply opposite conditions predominates, the results of its action usually divide into the external and the internal. (BT, pp. 690–691)

Beelzebub then goes on to describe the formation of the stars and planets as a result of the action of the 'Emanation of the Sun Absolute' or 'Word-God' on the space of the Universe, combined with the altered functioning of the intervals in the law of seven (broken symmetry). But then:

> At this stage in the process of the first outer cycle of the fundamental sacred Heptaparaparshinokh, that is, after the formation of the third-order suns or 'planets,' owing to the changed fifth deflection of the sacred Heptaparaparshinokh—which, as I have already said, is called 'harnel-aoot'—just here, the initial impulse given for the completing process, having lost half the force of its vivifyingness, manifested only half of its action outside itself, keeping the other half for itself—that is, for its own functioning—

in consequence of which, on these last big results called 'third-order suns' or 'planets,' there began to arise what are called 'similarities to the already arisen.' (BT, p. 694)

These 'similarities to the already arisen' consist of 'microcosmoses,' or single cells, and 'aggregations of microcosmoses,' or plants and animals. These are considered to be relatively independent cosmoses in their own right, resembling the larger cosmoses in structure. So, because of the disharmonization of the interval between sol and la, at a certain point in the creation of the now existing universe, smaller cosmoses appeared on larger ones, the planets. This is somewhat similar, although in a sense reversed, to the idea that residual forces between subatomic particles lead to atomic forces and residual forces between atoms lead to molecular forces. In both cases, broken symmetry leads to the creation of new cosmoses.

* * *

The do-sol harmony, the fifth, 3:2, is the most harmonious sounding after the octave, giving sol a special place in the scale. As indicated by its placement in the enneagram, the sol-la interval can be seen to provide extra force for the completion of the octave and the passage from si to do. This could also be related to the fact that it is the last superparticular ratio in the scale, before the octave. The Pythagoreans made many distinctions among ratios of different kinds, one of which is the singling out of ratios in which the larger number contains the smaller number and only one "part" of it, which ratios they called superparticular (Taylor, pp. 41–44). The formula for such a ratio is $1+1/n$. Thus, in the ratio 3:2, the 3 contains 2 (the n) and another half of 2, and the ratio can be written as $1+\frac{1}{2}$. In the ratio 9:8, the 9 contains 8 and another eighth of 8 ($1+\frac{1}{8}$). All of the first four ratios in the scale are superparticular: $\frac{9}{8}, \frac{5}{4}, \frac{4}{3}, \frac{3}{2}$. However 5:3 and 15:8 are not superparticular, as the larger number contains the smaller and several parts of it: in the ratio 5:3, 5 contains 3 and two thirds of 3. These were called superpartient by the Pythagoreans. The only remaining superparticular ratio in the scale is the octave, 2:1 ($1+\frac{1}{1}$), although it was considered special as it is a doubling. Still the sequence 5:4, 4:3, 3:2, and 2:1 could be considered a natural one within the octave and relate to the impetus provided by the interval sol-la.

Sol is also the first note on the left side of the enneagram. As will be discussed below, the right side of the enneagram can be considered to be related to the outer world, and the left side to the inner world. This corresponds also to a change in level, or cosmos. So, for a variety of reasons, sol is a pivotal note within the octave.

9

Eight, nine, and the symmetry groups of the enneagram

The number 8 also has a special place in physics, as in the octet of elementary particles:

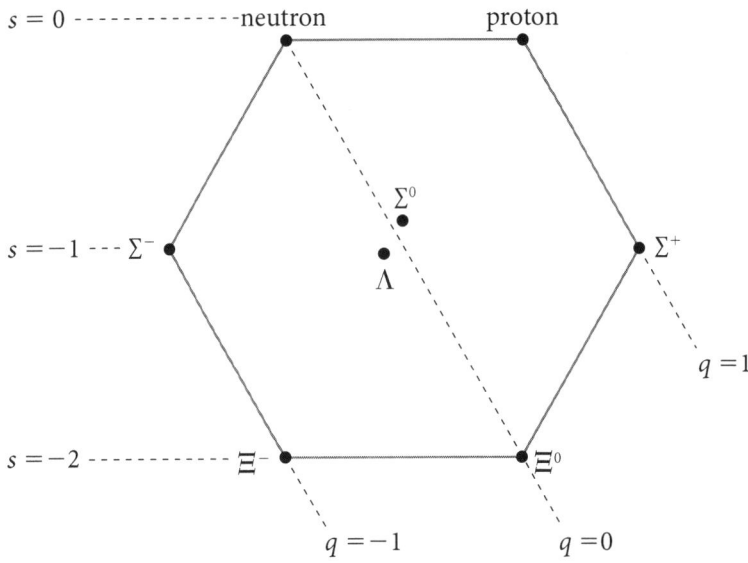

Possibly more importantly, eight is the number of the octonions, about which more will be said later.

The scale with its 'shocks' can be put into a symmetrical pattern of eight notes (GET, p. 54):

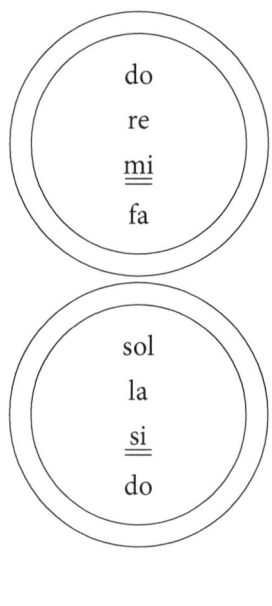

*   *   *

In order to fully appreciate some of the subtleties encrypted in the enneagram, it is necessary to look at some of the mathematics of group theory and abstract algebra, particularly as they pertain to modular arithmetic. Group theory has been essential in the development of modern physics, and its application to the numbers of the enneagram is intriguing.

A group, in mathematics, consists of a set of things (often these "things" are numbers, but they can be anything, including subatomic relationships) along with an operation that tells how to "add", or "multiply", any one of these things in the group by any other one to get a third, which obeys certain rules. These are closure, associativity, the existence of an identity element, and the existence of inverses. These abstract and general rules are easiest to grasp with an example of a group. The integers (...−3, −2, −1, 0, 1, 2, 3...., from minus infinity to plus infinity) form a group under the operation of ordinary addition. If two integers are added, we always get another integer (closure). If we add two integers and then add a third to the result, it doesn't matter in which order the additions are done: 3+(2+4)—first add 2 and 4, then add 3 to the result—is equal to (3+2)+4—first add 3 and 2, then add the result to 4 (associativity). There is an identity element, 0, which added to any integer leaves it unchanged: 3+0=3=0+3. Finally, every integer has an additive inverse in the set, which when added to it gives the identity element. So, (-3)+3=0, 5+(-5)=0. While for the integers the result of adding two numbers together does not depend on the order in which they are added (3+2=2+3), this is not the case for every group.

By this set of rules, the integers do *not* form a group under the operation of multiplication. There is closure (two integers multiplied together always gives another integer), associativity (3×(2×4)=(3×2)×4), and an identity element, 1: (1×3=3=3×1). However, most elements in the set of integers do not have a multiplicative inverse in the set because the inverse of, say 4, would be ¼ (4×¼=1) and fractions are not in the set of integers. So, to have a group under multiplication, the set has to be expanded to include the rational numbers (fractions), excluding 0—because 0 cannot be multiplied by anything to give 1.

Groups are the mathematician's way of dealing with symmetry, because the set of objects that form a group can be thought of as a collection of symmetries, in the sense that the group operation leaves something unchanged (the new element is still in the set) while something else is changed (it is a different element of the set). For instance, the rotations and reflections that leave an equilateral triangle unchanged *in appearance* form a group. The rotations and reflections are the set of "objects", and the operation is combining any two of them. So, a combination of two rotations of 120° results in another member of the set, namely a rotation of 240°. Two successive reflections about the same axis leave the triangle unchanged (completely unchanged, not just in appearance), "no change" being the identity element, so the two reflections are inverses of each other. Similarly a 120° rotation is the inverse of a 240° rotation, since applying them in succession leaves the triangle unchanged. It is easily shown that the result of any succession of rotations and/or reflections can be obtained by a single rotation or reflection, so there is closure.

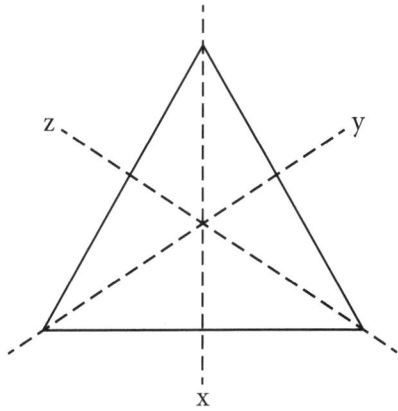

The group of integers under addition also represents a symmetry: the operation of addition always gives an integer, albeit a different one. What has changed is the integer; what has not is that it is an integer, not a cat.

The positive integers modulo 9 form a group under addition. The set consists of the integers 0–8, or alternatively 1–9, 9 being the same as 0. All positive integers can be reduced to a single integer between 0 and 8 by modular or "theosophical" addition: 16 becomes 7, 56 becomes 11 and then 2, etc. Confining ourselves to the integers of the enneagram, 1–9, adding any two gives another one (closure), e.g., 8+8=16=7 modulo 9. Associativity is preserved. Zero, or 9, is the identity: 0+any integer = that integer; and there is an additive inverse for each number 1 through 9: 1+8=9=0, 2+7=9=0, etc.

There is a multiplicative identity, namely 1, because 1 × any integer = that integer. However, the positive integers modulo 9 do *not* form a group under multiplication because some of the numbers 1 to 9—3, 6, and 9—cannot be multiplied by any other member of the set to give 1. But the remaining integers 1, 2, 4, 5, 7, and 8 do have multiplicative inverses in this subset. This subset is also closed under multiplication, as can be seen below; the multiplicative inverses are indicated in bold.

1×1=1	2×2=4	4×5=20=2
1×2=2	2×4=8	**4×7=28=1**
1×4=4	**2×5=10=1**	4×8=32=5
1×5=5	2×7=14=5	5×7=35=8
1×7=7	2×8=16=7	5×8=40=4
1×8=8		7×8=56=11=2
		8×8=64=10=1

Thus, the subset of numbers 142857 have multiplicative inverses and form a group under multiplication. This subset is no longer a group under addition, as there is no 0. Mathematically, the numbers 124578 form a group under multiplication modulo 9 because each of the digits is relatively prime to 9; that is, they have no common divisor with 9.

The numbers 3, 6, and 9 have additive inverses, within their subset, and form a subgroup under addition: 3+3=6, 3+6=9=0, 3+9=12=3, 6+6=12=3, 6+9=15=6, 9+9=18=9=0.

The numbers 3, 6, and 9 do not form a group under multiplication because there is no 1, and because they are not relatively prime to 9; but multiplication of the numbers in this subset always gives 9(0): 3×3=9, 3×6=18=9, 3×9=27=9, 6×6=36=9, 6×9=54=9, 9×9=81=9.

So, in this system of the positive integers modulo 9, the three parts of the enneagram are naturally segregated in yet another way: 1-9 (or equivalently 0-8) form a group under addition only; 142857 form a group under multiplication, but not under addition; and 369 form a group under addition and have a distinctive behavior under multiplication.

Just as the integers form a group under addition but not under multiplication, and the addition of rational numbers (fractions) is required to form a group under multiplication, the integers modulo 9 form a group under addition only, and for the formation of a group under multiplication, only the digits of the fraction $\frac{1}{7}$ can be used. If one arranges the numbers 142857 with their multiplicative inverses (which can also be called their reciprocals), one obtains:

$$1 \longrightarrow 1$$
$$2 \longrightarrow 5$$
$$4 \longrightarrow 7$$
$$8 \longrightarrow 8$$

So, there is an interesting parallel: the digits 1-9 modulo 9 behave like the integers, and the digits 142857 modulo 9 like the rational numbers (excluding 0).

*   *   *

The chromatic scale (see pages 46 and 56) forms a cyclic additive group, modulo 12, the notes being the elements of the set, and the operation movement up the scale. Notes an octave apart are considered to be the same note. Any movement up the scale results in landing on another note of the scale (closure). There is an additive identity, 0 or no movement, and each movement has an inverse: a fifth (seven semitones up) followed by a fourth (five semitones up) results in an octave (12 semitones up, and 12=0). Associativity is evident. The group is called cyclic because all of its elements can be generated by repetitive application of a single operation. Thus, successive movements of a semitone will clearly generate the entire scale. Movements of 11 notes will have the effect of going down the scale: do to si, si to la#, la# to la, etc. These two generators are inverses of each other. The only generators of a cyclic group are those that are relatively prime to the modulus, meaning that the generator and the modulus have no factors in common (other tan 1). In the case of modulus 12, the numbers 1, 5, 7, and 11 are the generators, because all the other numbers from 1 to 11 have at least one factor in common with 12: 2 divides itself and 12; 6 shares factors 2, 3, and 6 with 12, etc. So, the note sol—the fifth, seven semitones up the scale—is a generator of the entire sequence, as is evident in the cycle of fifths (p. 56), ignoring the Pythagorean comma. Similarly, the note fa—the fourth, five semitones up, which is the additive inverse of sol—generates all the other notes. This suggests yet another reason for the importance of the note sol in the scale (see pp. 56–60).

In the cyclic additive group of the enneagram (modulo 9), whose elements are the numbers 1–9, the generators of the entire group are 1, 4, 2, 8, 5, and 7, all relatively prime to 9. Thus, for instance: 4+4=8; 8+4=12=3; 12+4=16=7; 16+4=20=2; 20+4=24=6; 24+4=28=1; 28+4=32=5; 32+4=36=9; and 36+4=40=4. The numbers 3, 6, and 9 can only generate themselves. So, the inner hexagram numbers generate the entire scale, and, perhaps fancifully, one could propose that, in parallel, the outer world of materiality is generated by the inner world of consciousness. (See next chapter.)

The two pairs of generators of the chromatic scale group add up to 12, being additive inverses: 1+11=12, 7+5=12. Similarly, the numbers 142857 come in three pairs of additive inverses modulo 9: 1 and 8, 2 and 7, 4 and 5, each pair adding up to 9, so that the entire subset 142857 adds up to 9 by theosophical addition.

It is interesting that the same requirement—relative prime status—exists for both the ability of the subset 142857 to form a multiplicative group, and for it to be the subset of generators of the entire original additive group.

10

The creation of the world and the three dimensions of time

How does all this relate to the world outside of abstract mathematics? Clearly there is a broken symmetry represented in the enneagram. According to Gurdjieff, the universe as we know it was created by the creator—by breaking the symmetry of the law of seven—because the 'abode' of the creator was gradually wearing down:

> ...our Omnipotent Creator once ascertained that the Sun Absolute, on which He dwelt with His cherubim and seraphim, was almost imperceptibly, yet steadily, diminishing in volume.
>
> As this fact ascertained by Him appeared very serious, He decided to review immediately all the laws which maintained the existence of that still unique cosmic concentration.
>
> During this review, it became clear to our Omnipotent Creator for the first time that the cause of this gradual diminishing of the volume of the Sun Absolute was simply the 'Heropass,' that is, the flow of Time itself. (BT, p. 685)

One cannot reasonably interpret the 'Sun Absolute' as referring to a star or anything grossly material. I think the best interpretation is that it refers to a universal consciousness, which was gradually diminishing because of entropy, or the gradual dispersal of order over time. Modern physics considers that entropy, or disorder, always increases in a global sense, though it may decrease locally and temporarily, as in an organism, which is a highly organized entity, as the word suggests. But an organism, according to physics, has to maintain its orderly structure at the expense of increased disorder outside of itself, and the overall result is an increase in entropy globally. The global increase in entropy over time is a necessary consequence of there being only one dimension, and direction, of time. In fact, the direction of time, from past to future, can be defined as the direction in which entropy increases.

The classic example of the increase in entropy in a physical system involves two containers of gas, one of which has a higher temperature than the other because the gas molecules are moving faster in that container. If the two containers are brought together and a door opened between them, they will gradually reach the same temperature, because the gas molecules will pass through the opening randomly, and the faster ones will have more chances to do so. As a result, and also because of collisions between molecules, little by little the average speed of the molecules will become the same in the two containers. The only antidote to this would be to have an agent at the opening—christened 'Maxwell's demon' by physicists—who could see the speed of the molecules approaching the opening from either side, and open and close the door in such a way that he would only allow slower molecules to pass into one container and faster ones to pass into the other. However, the perceptions and actions of the demon would need energy from somewhere outside the system of the two containers, and this energy use would cause an increase in entropy overall. A living organism does the same thing: using energy from outside, it maintains a selective order inside itself. If the organism died, its molecules would disperse; so it must actively maintain its internal order. There is a local decrease in entropy, but because it is using energy from outside, by eating and breathing, perceiving and acting, overall there is an increase in entropy in the world.

The only way out of this conundrum is if there is a possibility of maintaining order without a greater increase in disorder elsewhere.

According to Gurdjieff, the purpose of the creation of the now existing universe was to counteract entropy, and if we regard the 'abode' of the creator as consciousness, it was to replenish, or maintain, consciousness. This was ultimately made possible through the evolution of conscious beings, who could contribute to the consciousness of the universe, becoming as brain cells in the mind of God. Referring to beings like us, called three-brained beings:

> I once told you that in them, as in us, the head is the place of concentration of cosmic substances, the total functioning of which corresponds exactly to the totality of functions that our Most Holy Protocosmos [God] fulfills for the entire Megalocosmos [the universe].
>
> This concentration of substances, localized in their head, they call the 'head brain.' The separate 'ossaniaki' or 'poptoplasts' of this localization or, as terrestrial learned beings call them, 'brain cells,' are destined to fulfill exactly the same purpose for the whole presence of each of them as is fulfilled for the whole of our Great Universe by the perfected highest bodies

of three-brained beings who have already united themselves with the Most Most Holy Sun Absolute or Protocosmos.

When these highest parts of three-brained beings, perfected to the corresponding gradation of Objective Reason, attain this union, they fulfill precisely that function of the 'ossaniaki' or 'cells of the head brain' which our Uni-Being Common Father Endlessness foresaw at the creation of the now existing world, when He graciously decided to use in the future those coatings that obtained independent individuality in the 'tetartocosmoses' as an aid for Himself in the administration of the enlarging world. (BT, p. 712)

'Tetartocosmoses' is another word for three-brained beings. Gurdjieff's idea is that three-brained beings are uniquely capable of developing souls that can manifest 'Objective Reason,' by which he means the highest degree of consciousness, or awareness of oneself and the world of which one is a part, possible for us. This possibility depends on having three brains. The third brain, or intellectual brain, is unique to three-brained beings. While other mammals certainly have what we call brains in their heads, including quite a bit of cerebral cortex, there is an obvious change of level in the capacity for thought in humans, manifest as language, abstract thought, mathematics, technology, etc. How the human brain made this quantum leap in capacity is a subject of controversy.

Since entropy is the result of the flow of ordinary time, one could regard its conquest as requiring another "dimension" of time. Gurdjieff suggested that each of the three brains is related to a different dimension of time, though he did not elaborate much:

> Time is the most important thing next to awareness. The flow of time through us gives us our chance to extract what we can. Time is a three-fold stream passing through our three centers. We fish in the stream, what we catch is ours, what we don't is gone. Time does not wait for us to catch all in the stream. If we catch enough we have enough to create the three bodies and become enduring. (Archive.org)

P.D. Ouspensky, a prominent early pupil of Gurdjieff's, had a more detailed theory of three dimensions of time, which Gurdjieff appeared to tacitly approve (ISM, pp. 208–213). Ouspensky's three dimensions of time were: 1) ordinary time, moving steadily from past to future, one second at a time, in which one of many possible sets of events are actualized each moment; 2) the dimension of "eternity," in which each moment in a sense exists forever;

and 3) "the line of the actualization of all possibilities." As discussed more fully elsewhere ("Does Man Have Three Brains?" In MiC, pp. 147–186), each of these three dimensions can be thought of as having a special relationship with one of the three brains. The body moves in ordinary time and cannot escape it, and its brain, centered in the spinal cord and brainstem, generates a wide variety of reflexes to keep it intact as circumstances change. The body always exists *now*. The mind, centered in the cerebral cortex, however, is free to look forward and backward in time and examine all possibilities; it gives us the capacity to remember, analyze, plan, and foresee. The feelings seem particularly connected with the dimension of eternity: feelings have a quality of being "forever," and they are strongly related to the perception of "energy" or "life." Energy in turn is closely related to vibration—the energy of a photon is proportional to the frequency of its associated electromagnetic wave—and vibration, being cyclical, a pattern that recurs over and over, has a quality of eternity. Gurdjieff regarded the feeling organization as centered in the autonomic nervous system, which in fact regulates all the energies of the body. Another, related, tripartite division of the world is into matter, which exists in ordinary time and is subject to disintegration by entropy; energy, which can change its form but never be destroyed; and "form," the abstract structure of things and relationships, which includes all possibilities. This division is currently described as matter, energy, and information, and corresponds to basic ideas in physics.

The enneagram has three parts, each of which can be related to one of the dimensions of time. The movement around the circle from re to do or from 1 to 9 represents movement and progression in ordinary time. The 142857 figure represents the looking forward and back, over multiple time scales, that the mind is capable of, and the triangle represents eternity.

> The isolated existence of a thing or phenomenon under examination is the closed circle of an eternally returning and uninterruptedly flowing process. The circle symbolizes this process. The separate points in the division of the circumference symbolize the steps of the process. The symbol as a whole is *do*, that is, something with an orderly and complete existence. It is a circle—a completed cycle. It is the *zero* of our decimal system. (ISM, p. 288)

Movement around the circle according to the notes of the scale is a step-by-step process of transformation, which also has a cyclical, or more correctly, spiral aspect. The body's transformations proceed in ordinary time, in a step-by-step progression. This corresponds to the group of integers modulo 9 under addition (whose identity element is 0). On the other hand, the mind can jump back and forth in time and possibility without going through intermediate steps. This corresponds to the group under multiplication formed by the numbers 142857 (whose identity element is 1), multiplication being a 'faster' process than addition, on a different level, in a sense. Interestingly, Gurdjieff makes a distinction between the progressive transformations of matter and the movements of intelligence:

> But, in this connection, it must be remembered that 'intelligence' is determined not by the density of matter but by the density of vibrations. The density of vibrations, however, increases not by doubling as in the octaves of 'hydrogens' but in an entirely different progression which many times outnumbers the first. (ISM, pp. 319–320)

(It is intriguing that in his lecture on symbolism and the enneagram [GET, pp. 49–71], having identified the circle of the enneagram as corresponding to zero [as in the quote on the previous page from ISM], Gurdjieff then precedes the list of the decimal expansions of the fractions $\frac{1}{7}$, $\frac{2}{7}$, etc., by this: "0 equals 1" [GET, p. 62]).

Intelligence, by remembering and anticipating, planning, arranging and correcting, is the ultimate antidote to entropy. J.G. Bennett, a student of Gurdjieff's, called the action of intelligence the "overcoming of hazard" (Bennett, pp. 16–21). In his book, he describes a variety of processes which are made possible by looking forward and backward according to the 142857 hexagram. For instance, in the process of preparing a meal, the cook has to look forward to what the raw ingredients will be in order to lay out the necessary kitchen tools, and, slightly further along, he has to anticipate the people who will eat the meal and the layout of the table where the meal will be eaten in order to cook the right things and amounts. Although one can make connections between almost anything and almost anything else, much of the way Bennett describes the interactions between the progressive process of making a meal and the intelligent movements forward and backward in time and possibility of the cook's awareness seems sensible and non-arbitrary.

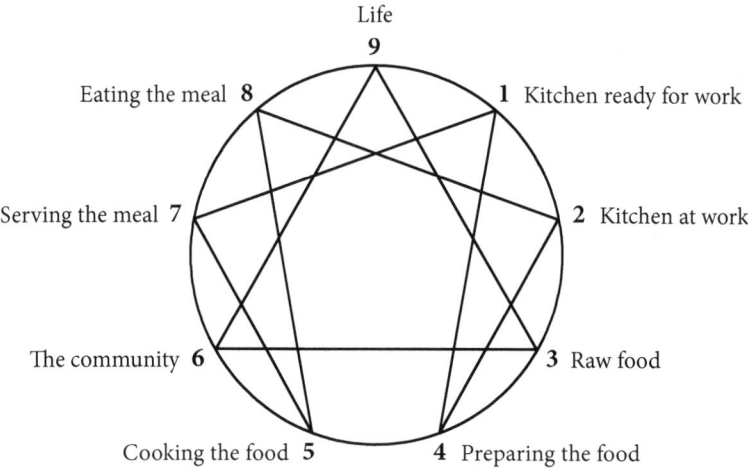

THE KITCHEN AS A COSMOS

Bennett, pp. 22–34

Blake, one of Bennett's students, has a particularly nice expression of the significance of the hexagram:

> The inner lines weave their cohesion of memories and anticipations, eternally present, expanding the range and depth of the "here and now." (Blake, p. 43)

In any case, the general principle is indisputable: physical processes cannot move along in time except stepwise, while intelligence has no such limits, which is how it can overcome entropy. Whether it can overcome entropy only locally, or whether, as Gurdjieff claims, the universe was created in such a way as to be truly self-sustaining, a real 'perpetual motion machine,' is a more difficult question. Recall that Gurdjieff described the enneagram as representing perpetual motion, but he also mocked humanity's attempts to create a mechanical perpetual motion machine (BT, pp. 69–71). At one time, trying to create a perpetual motion machine was quite a fad—my high school science department had a display of such attempts—but none of these attempts were successful because friction inevitably wastes some of the energy, making the machine fail eventually. Friction in this situation corresponds to entropy. Perpetual motion is only possible if consciousness is part of the picture—in part because consciousness can use friction productively (see p. 85)—and requires more than one dimension of time. The inexorable increase in entropy in the universe claimed by classical physics could be seen as a result of two related deficiencies in its worldview: the virtual exclusion of consciousness and an incomplete understanding of time.

It has been evident since the advent of quantum mechanics and relativity theory that the fundamental patterns that underlie the construction of the world are not what we imagine them to be based on our interactions with terrestrial materiality. We imagine that atoms are "little things" that aggregate to form bigger things, just like grains of sand in a sandpile. Some very complicated bigger things, like animals, are able to interact with other things and even be aware of them. But the fundamental constituents of matter are not describable as things, or even miniature solar systems—are they waves, or particles, or both, or neither? Mathematically, they are vibrations but not vibrations of anything substantial; rather they are vibrations of possibility (probability amplitudes). Since atoms and subatomic particles also persist in ordinary time, all three dimensions of time are represented: possibility, eternity—as vibration—, and ordinary time. So, atoms might be regarded not so much as little "things" but rather as little bundles of three-dimensional time.

11
The triangle and eternity—3, 6, and 9

The triangle has a number of symbolic meanings. It "connects together the law of seven and the law of three" (ISM, p. 290). The law of seven is the law that governs processes, and the law of three is necessary in any interaction. So there is a law of three that applies at each note of the octave, or each stage of the process, but also a law of three that governs the entire octave. It represents the 'higher', or 'life'—energy—entering into the process.

> When life is extinct—for instance, if a plant is cut down—the energy inherent in it is dispersed. The triangle comes out of the symbol of the octave of life. (VRW, p. 221)

> The triangle is not a process but a structure of intention or will. It is what is *informing* the process and giving it meaning. (Blake, p. 25)

It is fitting then that the triangle includes the beginning note and also the two intervals where "outside" help is needed to keep the process from deviating. The triangle also represents unity in multiplicity, which Gurdjieff regarded as of paramount importance:

> One of the most central of the ideas of objective knowledge is the idea of the unity of everything, of unity in diversity. (ISM, p. 278)

> The apex of the triangle closes the duality of its base, making possible the manifold forms of its manifestation in the most diverse triangles, in the same way as the point of the apex of the triangle multiplies itself infinitely in the line of its base. (ISM, p. 288)

These multiple meanings of the triangle, as representing the life of a thing or process, the higher, unity in multiplicity, and also the dependence of processes on interactions with the world outside, indicate that the horizontal relationships between entities or processes and the vertical relationship of entities with the higher or God are two aspects of the same thing. This is also indicated by the 'cells of the head brain' analogy, referred to on pp. 68–69. For the cells of our brain are only useful as a result of their interactions with other cells. For instance, any given photoreceptor cell in the retina in the back of the eye can only indicate the presence or absence of light at that location. In order for the shape of objects to be perceived—the first step in making visual sense of the world—many photoreceptors must compare notes, determining the boundaries between light and dark and different colors. Much of the brain functions in this way, through multiple levels of comparisons between cells, ultimately making possible our capacity for abstraction and analogy, and thus our ability to perceive unity in multiplicity.

One could even say that God is not separate from the universe, but *is*, in some aspect at least, the conscious universe, which in turn is sustained by the mutual awareness of individual consciousnesses. This idea is present in many traditions.

> In the Heaven of Indra, there is said to be a network of pearls, so arranged that if you look at one you see all the others reflected in it. In the same way each object in the world is not merely itself but involves every other object and in fact IS everything else. (Eliot)

> Knowing that the individual Self, eater of the fruit of action, is the universal Self, maker of past and future, [the wise man] knows he has nothing to fear.
> Born in the beginning from meditation, born from the waters, having entered the secret place of the heart, He looks forth through beings. That is Self. (Katha Upanishad)

> Only he may enter here who enters into the position of the other results of my labors. (BT, p. 1065)

> There is obviously only one alternative, namely the unification of minds or consciousnesses. Their multiplicity is only apparent, in truth, there is only one mind. This is the doctrine of the Upanishads. And not only of the Upanishads. The mystically experienced union with God regularly entails this attitude unless it is opposed by strong existing prejudices. (Schroedinger, p. 85)

> Jesus said unto him, Thou shalt love the Lord thy God with all thy heart, and with all thy soul, and with all thy mind. This is the first and great commandment. And the second is like unto it. Thou shalt love thy neighbor as thyself. On these two commandments hang all the law and the prophets. (Matthew 22, 37–40)

More on the triangle, and the mingling of dimensions

If time is regarded as three dimensional, then those dimensions, like the three spatial dimensions, must in some sense be perpendicular to each other. However, the choice of the three directions in space is arbitrary, as long as they are perpendicular, and things change their appearance depending on the viewpoint of the observer. For instance, a circle, as seen from above, becomes a line when viewed from the side, and an ellipse from intermediate angles. In fact, the search for the fundamental laws of physics is the search for what does *not* change when the viewpoint—in the broadest and most abstract sense—is changed, hence the importance of group theory.

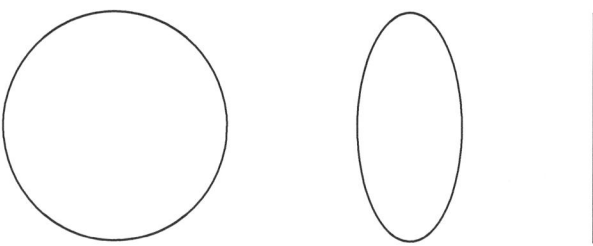

The dimensions of time also change their appearance depending on the viewpoint. The perpendicularity of the dimensions of time can be considered to be symbolized in the enneagram by the fact that the circle and the triangle are two perpendicular views of a cone.

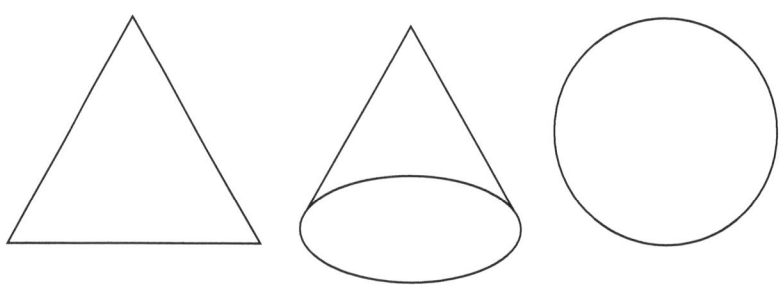

A number of interesting transformations are observable, corresponding to the way that the different aspects of time can change into each other. The triangle, seen as a cone, contains a progressively larger series of circles emanating from its apex, corresponding to the progressive increase in wavelength and decrease in frequency of the vibrations emanating from the creator.

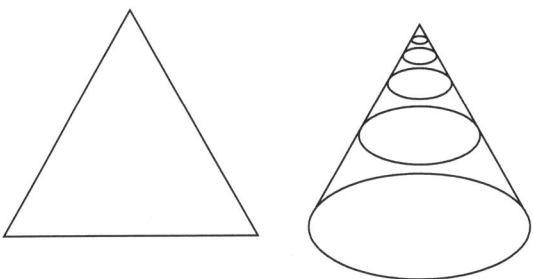

For a sine wave, the basic pattern of vibration, is movement along a circle, projected onto a perpendicular moving plane:

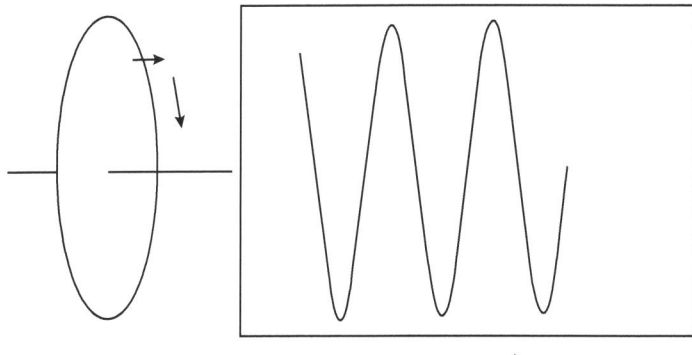

The triangle, representing the dimension of eternity, thus is intimately connected with vibration and with cyclic phenomena. As mentioned above, the emotional brain is intimately related to the energies of the body, to its "life." The autonomic nervous system, which Gurdjieff equates with the emotional brain (BT, pp. 713–714), regulates all the energies of the body.

But movement in ordinary time, along the outer circle of the enneagram, becomes cyclic once the octave is complete, thus also corresponding to a vibration.

The dimension of possibilities can be symbolized by a tree. Trees, whether they be of the plant variety, or family trees, or decision trees, represent the elaboration of possibilities in time. One pattern that governs growth can be seen in the sneezewort plant, whose leaves follow a Fibonacci rule, as described in chapter 5. Below, the arrangement has been changed slightly, without disobeying the rules, to produce a major scale at level 8.

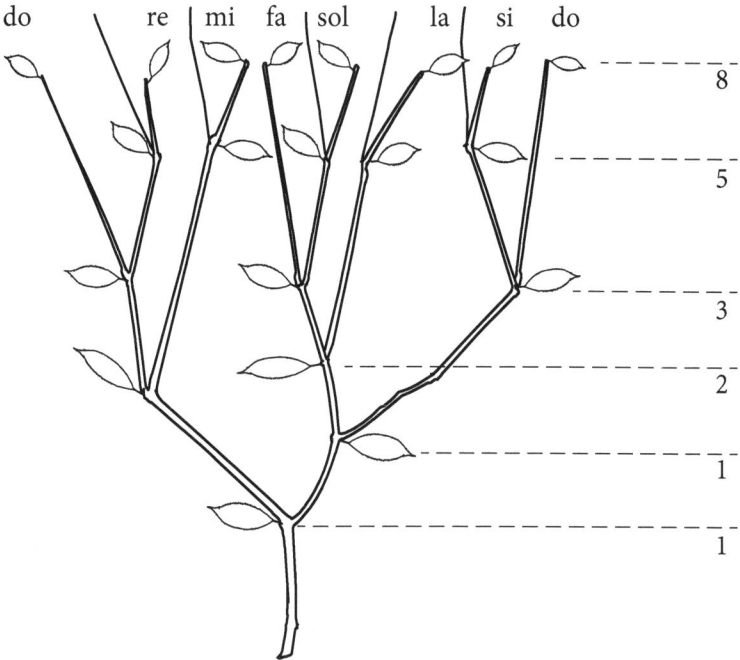

At the level where there are eight leaves, there are also five branches that pass through this level without elaborating a leaf, making a total of thirteen, which is the next Fibonacci number. This can be seen as the seven notes of the octave (the leaves) and the five intervening semitones (the branches), plus the final do that completes the octave. Thus an elaboration of possibilities, from another viewpoint, becomes a progression of notes, representing a process, which when complete forms a cycle, a vibration.

* * *

The hexagram of the enneagram, representing the dimension of possibility and the abstracting and unifying ability of the mind, can be related to the points and lines of projective geometry. Projective geometry largely grew out of the studies of perspective by Renaissance artists, who were seeking to accurately represent a scene stretching out horizontally in space on a vertical canvas (inherently involving a shift in dimension). To do this, one draws imaginary lines between the features of the scene and the artist's eye, passing through the canvas in front of the artist so that what is projected onto the canvas is exactly what the eye sees.

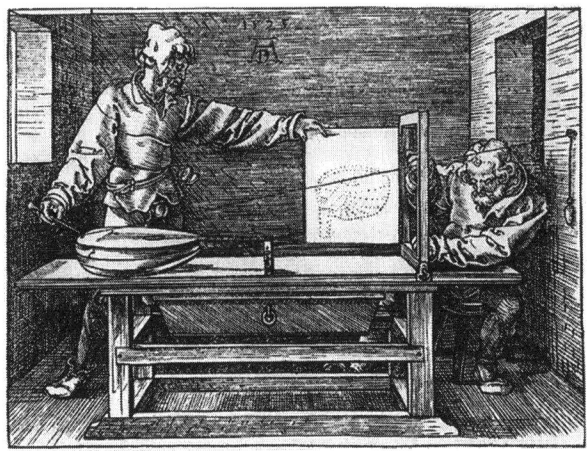

Durer woodcut illustrating perspective projection.

As a result, the actual geometry of the scene, as it would be depicted "from above," is distorted. Parallel lines get closer and closer with distance from the observer and would meet at an infinitely distant horizon. Equal distances appear smaller the further away they are.

Essentially the only features of the "actual" scene that remain intact are points and the straight lines that connect them. However a great deal of mathematical theory can be built on this "bare bones" foundation, which then finds application in more fleshed out geometries, such as the Euclidean geometry everyone learns in school.

A large part of the brain is devoted to dealing with perspective transformations. This is most obvious in the case of visual-motor coordination. Every time we move or turn in the world there is a different view of it, yet we effortlessly move around without falling or bumping into things much and can look, reach, and grasp accurately while in motion, not to mention the remarkable feats of coordination that characterize sports. All this requires a tremendous amount of neural computation, and large parts of the brain are devoted to it. Visual-motor coordination goes on largely subconsciously, but our conscious visual awareness requires similar calculations and transformations: as we look around, each time our eyes rest momentarily, there is a different view of the surrounding scene, yet in our minds, it is all pieced together into one (temporarily) stable mental image of the part of the world we are in.

Another aspect of projective geometry is that some things that are distinct in our habitual Euclidean space become the same. This is perhaps most dramatic in the case of the conic sections—the hyperbola, parabola, ellipse, and circle that are obtained by slicing a cone in different ways.

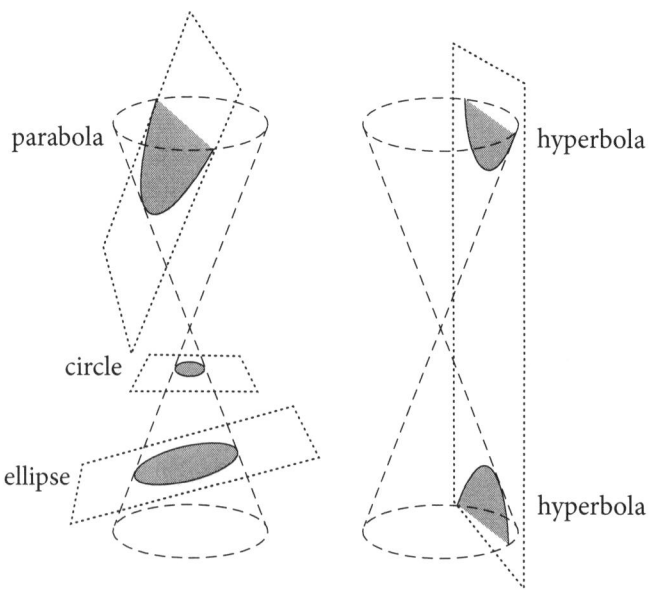

Under perspective transformations, all these curves can become the same. Everyone can experience this in the case of a circle becoming an ellipse, as shown page 76. It is more surprising to see a parabola, whose arms, looked at head on, become further and further apart, turn into an ellipse in a perspective rendering, because at infinity the two arms appear to come together.

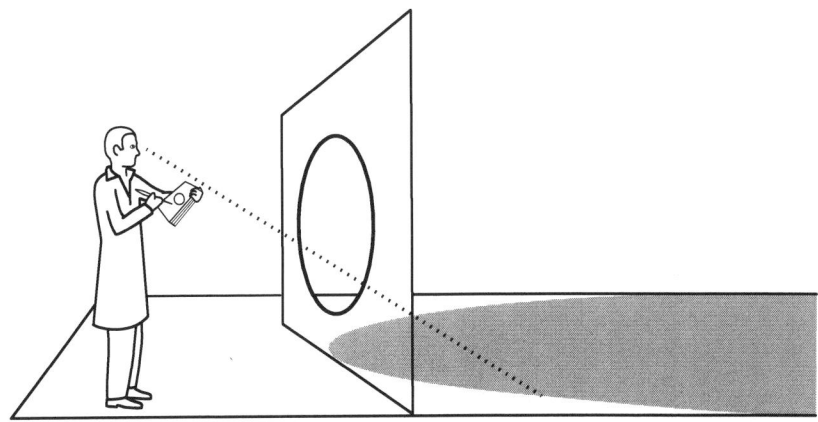

So, in a sense, the general conic section is an abstraction of the various specific conics, just as the number 2 is an abstraction of the various pairs of things one can encounter. This process of abstraction, in its various guises, is the main job of the human brain (third brain of Gurdjieff). As a result, we can move seamlessly in our minds between one place and another, between future and past, between one idea and another, and between one possibility and another, without passing through intermediate steps in physical, temporal, or conceptual space. All this is symbolized by the lines and points of projective geometry, and of the hexagram within the enneagram, and represents the third dimension of time.

Projective geometry, as its origin in perspective drawing shows, inherently requires an observer, whose eye is the focus of the projections. This is a two-way phenomenon: the eye receives the particular perception its location dictates, and at the same time, the eye creates the reality it perceives, for the perceiver. This is quite reminiscent of the paradoxes of quantum mechanics, in which the nature of a phenomenon (particle/wave?) depends on the observation being made. What is reality, then? As suggested above, perhaps it is intimately related to the combined perceptions of mutually aware consciousnesses. This is not to say that the moon disappears when I don't look at it, an absurdity that some have used to mock the idea that there is a connection between quantum uncertainties and consciousness. This is a much more subtle and elusive proposition, seemingly endorsed by Gurdjieff, and summed up in the phrase "God made man so that He could know Himself" ("The Ego and the I." In MiC, pp. 91–94), which, if understood scientifically, would drastically alter science's worldview.

One thing is clear: there is a strong resemblance between the observation's role in determining the nature of a phenomenon in quantum mechanics and the mind's role in deciding what is there when presented with ambiguous stimuli (which ultimately all stimuli are). The Necker cube is a particularly striking example of this: it can be seen in one of two ways and is always seen in one of them, but never both at once.

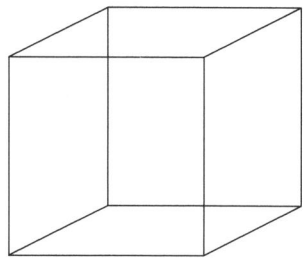

13

The mirror symmetry of the enneagram

One of the symmetries of the enneagram is a bilateral symmetry, or mirror symmetry, about a vertical line passing through point 9. This is similar to the external symmetry exhibited by many animals, including humans, and by the human brain. However, just as the body's external symmetry does not extend to many of the organs within it, and the human brain's external symmetry obscures fundamental differences in functioning between the two cerebral hemispheres, the enneagram's bilateral symmetry actually dictates an asymmetry of function of the inner hexagram. On the right side, the progression is from the bottom up: 1 has to go to 4 to then reach 2, passing by 3 each way. On the left side, the progression is from the top down: 8 to 5 to 7, passing by 6. It is of interest in this regard that Gurdjieff made a distinction between the mode of functioning of the two 'intervals,' represented at 3 and 6 in the enneagram, or between mi and fa and si and do in the scale. In the creation myth described in BT, the Creator is said to have lengthened the first interval and shortened the second (a difference from the description in ISM, where both intervals are said to be shorter).

> To provide the stopinder between the third and fourth 'points of deflection' with the required property of absorbing for its functioning the automatic inflow of all surrounding forces, He prolonged its duration.
>
> And this is the stopinder of the sacred Heptaparaparshinokh which is still called the 'mechano-coinciding mdnel-in.'
>
> And the stopinder that He shortened is between the last 'point of deflection' and the beginning of a new cycle of the completing process of this sacred law. By this shortening, in order to facilitate the beginning of a new cycle, He predetermined that the functioning of this stopinder would depend solely upon the influx through that stopinder of external forces resulting from the action of that cosmic concentration itself in which the completing process of this primordial sacred law is flowing.
>
> And this stopinder of the sacred Heptaparaparshinokh is the one that is still called the 'intentionally actualized mdnel-in.' (BT, p. 690)

The right side of the enneagram thus represents mechanicality, or the body's reflexes, or the first dimension of time, while the left side represents the top-down action of intention, or awareness, or the third dimension of time. Correspondingly, in the representation on the enneagram of the first stages of the digestion of the three kinds of food, the interval mi-fa is filled by air, provided by the automatic breathing of the body. The movement from 1 to 4 to 2 helps this happen, though how this corresponds to known physiology is unclear. The interval sol-la requires the intentional act of self-awareness, which both helps the passage through the mi-fa interval of the air octave and allows the impressions octave to get started.

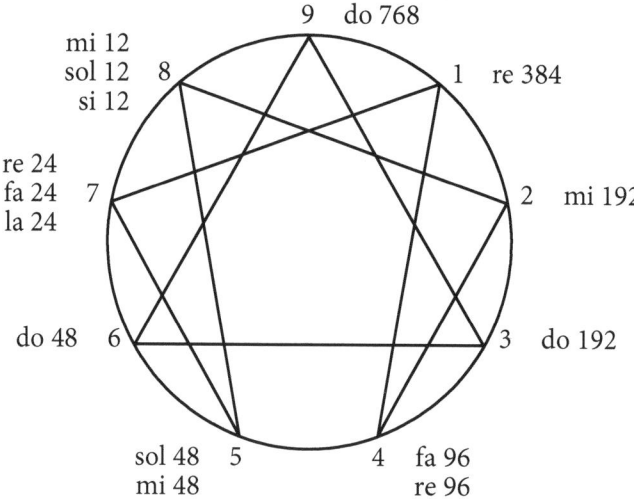

The triangle, representing the second dimension of time—eternity, energy and meaning—joins the two sides together. Perhaps the notes fa and sol, under the base of the triangle, are particularly related to this dimension also. Thus we have two further representations of the three dimensions of time: right side, left side, triangle, and 1–2, 4–5, 7–8. Multiple interacting representations give the enneagram its power as a symbol.

Gurdjieff called the enneagram a perpetual motion machine because it reflects the basic laws of the universe, which he also regarded, once the changes in the laws were made, as a perpetual motion machine. Prior to the change in the laws, there was only one dimension and direction of time, that of ever increasing entropy, and the Sun Absolute on which 'His Endlessness' dwelt was dwindling. Current physics also has only one dimension of time, which may be why it cannot encompass consciousness, and why the paradoxes of quantum mechanics are so inscrutable. The changes in the law of seven resulted, after a long process of creation and evolution, in a new dimension of time, corresponding to conscious intelligences, which have the power to foresee the future, remember the past, examine all possibilities, and act constructively. So now there is a perpetual struggle, between automaticity and intention, disorder and order, broken and restored symmetry, in the universe and in man, the three-brained being whose structure exactly mirrors that of the universe. This is reflected in the two sides of the enneagram. In us the struggle is between the higher nature of consciousness, conscience, intention and participation, and the lower animal nature whose only imperative is self-preservation. There are now two kinds of causation, corresponding to two dimensions of time. As Blake put it (p. 50): "Causal change depends on what has already happened. Intentional change depends on what *will* happen."

The result of this struggle is the perpetual motion of the universe, and a corresponding possibility of immortality for man. This perpetual motion creates, and was created by, the triangle in the enneagram, the eternal now, the third time dimension. The triangle reaches down to both sides, the automatic and the intentional. Both come from above. The automatic is a necessary part of the creation; the intentional comes from the call of universal consciousness.

In *Beelzebub's Tales to his Grandson,* Gurdjieff describes the new spaceship propulsion system (BT, pp. 66–68), which can be summarized as follows: instead of trying to blast away frictional resistance, using energy, the resistance is taken into the body of the ship and expanded, and then expelled out the back of the ship, propelling the ship forward. This chapter is immediately followed by the one in which Beelzebub mocks the earthlings' attempts to create a perpetual motion machine that is strictly mechanical. There is little doubt that he is describing the way in which the universe maintains itself, by a perpetual struggle between the active and passive forces, between consciousness and automaticity, intention and inertia, purpose and distraction. Three-brained beings are an integral part of this struggle, and it also gives them their own possibility of perpetual motion. The struggle involves taking in and meeting the resistance, not trying to get rid of it.

> With "I am" you begin to feel the two natures, their terrific battles. It is the perpetual battle of the world, of the two Principles: the Good and the Bad; in other words of the positive, active principle, and of the negative. It's the most terrible of all wars. It is also in us. You must see it. Between these two principles which are our very nature, you must bring to birth an individual: the

Man who is a measure, who has the power to make these two forces serve one Aim, the Man who acts for a Reason. (From notes reporting Gurdjieff's words at a meeting in the 1940s)

Time is the most important thing next to awareness. The flow of time through us gives us our chance to extract what we can. Time is a three-fold stream passing through our three centers. We fish in the stream, what we catch is ours, what we don't is gone.... If we catch enough we have enough to create the three bodies and become enduring.

Time is the sum of our potential experiences, the totality of our possible experiences. We live our experiences successively; this is the first dimension of Time. To be able to live experiences simultaneously is adding another or second dimension to Time. To be aware of this simultaneity is called solid time, or the third dimension of Time. When we have identified ourselves with Time it will be as Revelation says: "And Time shall be no more." (Archive.org).

14

The four normed division algebras

The possibility of multiple time dimensions appears intriguingly in the mathematical subject of the normed division algebras. There are four so-called normed division algebras (loosely speaking, algebras in which division is possible and there is a way to determine distance or magnitude), corresponding to four number systems: the real numbers (the ones we are familiar with, including integers, rational, and irrational numbers), the complex numbers, the quaternions, and the octonions. The complex numbers were created, or discovered, as a result of the fact that equations can have solutions that are square roots of negative numbers, and such square roots are not part of the real number system, since all squares, of both positive and negative numbers, are positive. So the square root of -1 was given the name i, for 'imaginary number.' The complex numbers, of the form a + bi (a and b being real numbers), require a plane for their geometric depiction, whereas for the real numbers a line is sufficient.

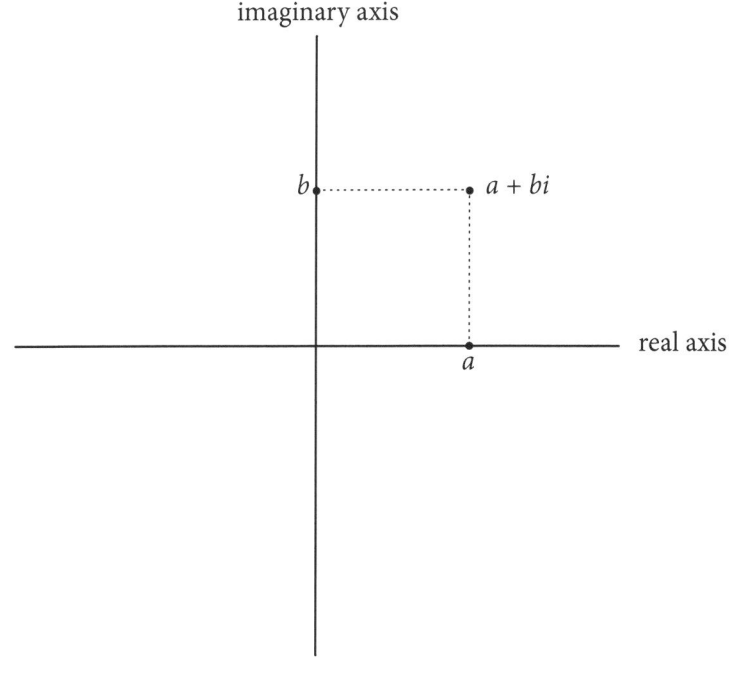

The complex numbers behave in ways different from ordinary numbers because of the square root of -1, i. Two complex numbers are added by simply adding their real and imaginary parts, e.g., (2 + 3i) + (4 + 5i) = 6 + 8i, but multiplying them produces peculiar behavior, because $i^2 = -1$:

$$(a+bi) \times (c+di) = (ac-bd) + (ad+bc)i$$

This results in a counterclockwise rotation, and expansion or shrinking, of the resultant vector; the original vectors' angles are added and their lengths multiplied.

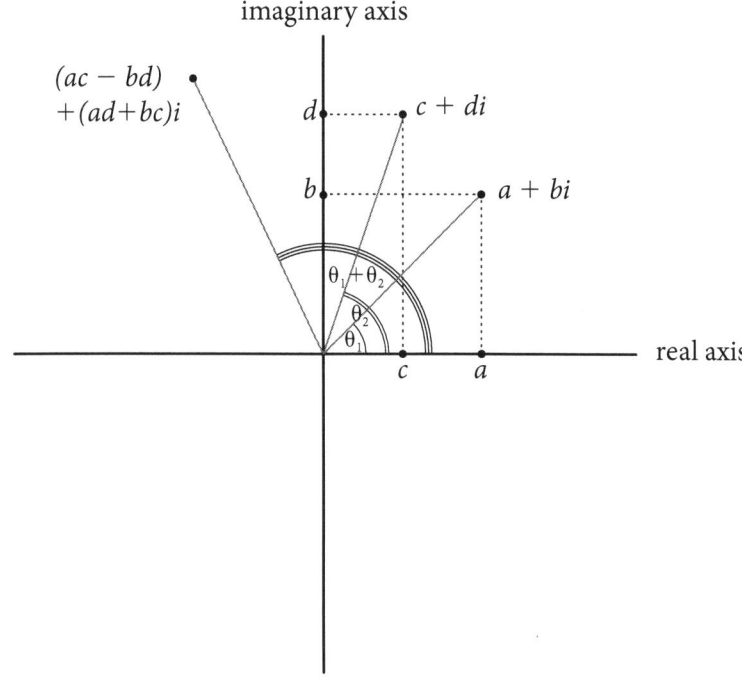

If one uses vectors of length 1, then multiplying them results in movement around a circle of radius 1, and this turns out to be the way to depict electromagnetic waves in quantum mechanics. Complex numbers are necessary for the mathematical equations to work correctly because the waves describe 'probability amplitudes,' rather than being "real" waves as are sound waves or water waves. So the complex plane is the field of the photon

Two further number systems are produced by doubling and then redoubling the complex numbers, resulting in the quaternions and octonions, respectively (Gray). The quaternions have one real part and three imaginary parts—three different square roots of -1, called i, j, and k, producing a four-dimensional space. The octonions have one real part and 7 imaginary parts, resulting in an eight-dimensional space. The imaginary octonion components are given various names and are depictable on a Fano plane, which we have encountered before (chapter 7), here shown with one of the nomenclatures. The arrows indicate the directions of the various multiplications of the imaginary components.

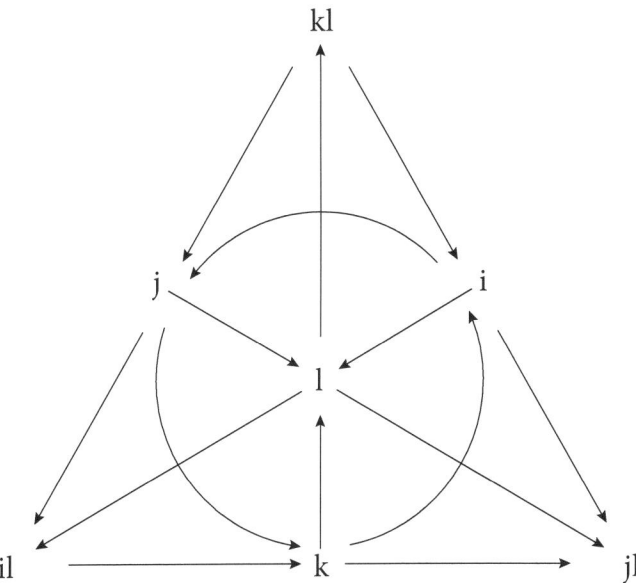

It turns out that just as the complex plane is the correct scaffolding for the symmetries governing the electromagnetic force, the quaternions can be used as the scaffolding for the symmetries governing the weak nuclear force, and the octonions for those governing the strong nuclear force (Gray).

A modification of the octonions, to make them 'split octonions,' in which some of the square roots of -1 are replaced by additional, now imaginary, square roots of +1, results in a space with four space dimensions and three time dimensions (Baez). This is tantalizingly similar to Ouspensky's scheme of three space and three time dimensions.

The Fano plane's octonions could be put on an enneagram in the following way, using the seven points of the enneagram corresponding to the notes of the scale:

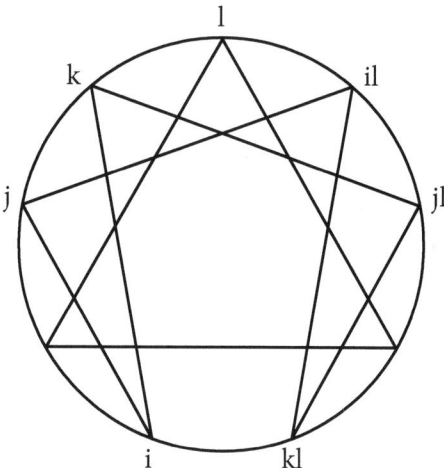

At this point I do not have enough mathematical knowledge to speculate further on what appear to be suggestive parallels between the patterns of the enneagram and these aspects of modern mathematics and physics. I have probably speculated too much already.

15
Fermions and bosons— space and time

Elementary particle-waves come in two distinct varieties: fermions and bosons. Fermions, with spin ½, obey the Pauli exclusion principle: no two can occupy the same space at the same time. They coalesce to form matter, which must therefore extend in space. The bosons, with spin 1, are not so restricted: any number of photons, or electromagnetic waves, swirl around my head simultaneously: light from all the objects around me, as well as radio and TV and cellphone emissions that appropriately tuned receptors could perceive. (The light we see with occupies one octave of frequencies in the electromagnetic spectrum; other, lower, frequency ranges are occupied by radio, TV, and cellphone emissions, while ultraviolet, x-rays, and gamma rays occupy frequency ranges higher than visible light.) Electromagnetic waves do not occupy space, rather they occupy frequency domains, or time.

Electromagnetic waves, or photons, unite everything. Electrons are maintained in their atomic "orbits" by a continual exchange of photons with the nucleus. The attraction and repulsion of charged fermions—such as electrons and protons—is mediated by photons. When atoms or molecules change their energetic state, it is by absorbing or emitting photons. Wherever molecules are vibrating, which is everywhere, electromagnetic waves are being produced. They are a universal medium of attraction and repulsion, transformation and exchange.

According to the theory of relativity, there is no space-time interval for a light ray between here and there (see also below). This is the light cone, which could be considered to represent the dimension of eternity, depicted by the triangle in the enneagram. Then the elaboration of frequencies within the cone—overtones, harmonies, beats, as described in chapter 12—makes up the dimension of possibilities.

There is increasing evidence that the forebrain operates to a significant degree by means of frequency relationships among its electromagnetic waves. These waves are generated by ensembles of cells whose membrane potentials are fluctuating relatively synchronously. It is thought by many neuroscientists that conscious unitary perception is mediated by harmonization of the rhythms of the brain areas that are involved in analyzing different aspects of the percept. There are multiple brain areas devoted to visual perception, for example, and different areas specialize in analyzing different aspects of visual stimuli, such as form, color, motion, and depth. Yet a conscious percept is unitary: all these aspects of a visual scene are seen together. At one time it was thought that all the components of a conscious perception are brought together in a place in the brain where consciousness resides, but no such place has been found. It is now increasingly evident that conscious perception comes together not in a place, but in time, by means of synchronized rhythms among the different brain areas involved. This gives new meaning to the concept of presence. It is likely that more inclusive levels of consciousness correspond to more global brain wave harmonizations.

It has also been shown that the brain wave rhythms tune themselves to attended stimuli. Patients with electrodes on the surface of their brains, for medical reasons, volunteered to watch a computer monitor on which on half of the screen a man was telling a story while on the other half a woman was telling a different story. The patients were asked, on any given trial, to pay attention to one of the stories and ignore the other. This has been called the "cocktail party problem" because at a cocktail party in which multiple conversations are going on, one can selectively attend to one of them, even if it is not the loudest. The patients' brain waves were recorded during these trials. Afterwards, the story that had been attended to on any given trial was predictable based on the rhythms of the brain waves, which reflected the speech rhythms of the attended speaker and not those of the other (Zion Golumbic).

> The reception of external impressions depends on the rhythm of the external stimulators of impressions and on the rhythm of the senses. (VRW, pp. 82–83)

Thus, brain rhythms become tuned not just to each other but to the rhythms of the external world as well.

So, the outer world of matter, of the fermions, is the world of space, while the inner world of consciousness, of interaction, and of the rejoining together by awareness of all that was separated in the process of creation, the world of the bosons, is the world of time. Each of these worlds is represented in multiple ways in the enneagram.

16

The inner and outer worlds, and the multiple intertwined aspects of the enneagram

The enneagram can be subdivided in a number of ways, corresponding to its symmetries. The movement around the circle can be equated with the outer life and the transformations of matter, while the inner hexagram corresponds to the inner life of the mind, which moves in a different way, along the dimension of possibility. It then makes a certain kind of sense that the outer movement involves a sequence of seven steps in order, while the inner movement is represented by the reciprocal, $1/7$ or 0.142857. For the outer and inner worlds are reciprocals of each other. From the point of view of the outer world, I, like any individual person, am nothing, a tiny speck on a tiny planet in a remote solar system, one of billions in a galaxy, which itself is one of billions. But from the point of view of the inner world, I am everything: everything I am aware of, perceive, know or remember—others, the immediate environment, the planet, solar system, galaxy, and universe—are in me, contained in my inner life.

> *The brain is wider than the sky*
> *For—put them side by side*
> *The one the other will contain*
> *With ease—and You—beside.*
>
> —Emily Dickinson

The reciprocal of the abundant profligacy of the creation of the outer world is the gathering back of all into a universal consciousness. Similarly, the modular arithmetic essential to the enneagram brings the infinite cornucopia of numbers back to their source, which, according to the Pythagoreans, is the first 10 numbers.

The body's trajectory in linear time is represented by the outer circle, the mind's movement in the world of possibilities by the hexagram, while the triangle represents the emotions' connection with life, energy, and eternity.
In calculating the space-time interval, or "distance" between two events, time

and space must be given opposite signs, as a consequence of Einstein's special relativity theory, so the Pythagorean formula for the interval in four-dimensional space-time becomes $\sqrt{t^2-x^2-y^2-z^2}$ (in purely three-dimensional spatial calculations it is $\sqrt{x^2+y^2+z^2}$). The opposite signs of the space and time components result in the zero overall space-time interval for a light ray between here and there. In the seven-dimensional space-time of the split octonions, space and time dimensions also have opposite signs. Taking the reciprocal in multiplication is equivalent to changing the sign in addition; for the additive inverse of a number (the one that when added gives 0) is its negative, and the multiplicative inverse (the one that when multiplied by gives 1) is the reciprocal. Correspondingly, in the outer world of physics, space and time have opposite signs, and outer processes are related to the additive group of the numbers 1–9, modulo 9, around the circle of the enneagram; whereas in relation to the inner world, represented by the hexagram, the number pattern results from the decimal expansion of $1/7$, whose numbers form a multiplicative group, modulo 9. (See chapter 9)

Interestingly, the reciprocal of i, the unit imaginary number so important in modern physics, is also its negative: $1/i = 1i/i^2 = i/-1 = -i$.

Numbers and their reciprocals have a sort of mirror symmetry, though it is a different kind of mirror than the ordinary one. The union (multiplication) of a number and its reciprocal results in the number 1, the multiplicative identity. Similarly, man's role in the universe is to unite the outer and inner worlds to form the 'third world of man' (Gurdjieff, *Life Is Real Only Then, When "I Am,"* pp. 143–177), which is the world of unity in multiplicity, symbolized by the number 1 and by the triangle of the enneagram. This involves a growth of the emotional part, which must evolve from self-concern to true consciousness and conscience, from isolation to participation, ultimately, according to Gurdjieff, resulting in the development of a soul that can participate in maintaining the consciousness of the universe.

One can also consider the right half of the enneagram to represent the mechanical movement of outer things, while the left half represents the conscious intentionality of a developed inner life. Then the mirror symmetry of 1-4-2 and 8-5-7 makes sense: mechanicality proceeds from below, while intention comes from above; mechanicality proceeds from past to future, while intentionality derives from the future goal.

Yet another subdivision relates the body to the numbers 1 and 2, the emotions to 4 and 5, and the intellect to 7 and 8, as suggested by the placement of the three foods on the enneagram.

The movement around the circle of the enneagram can proceed in either direction. The second do is an octave away from the first do, both at point 9, so the diagram really represents a spiral rather than a circle. It can be a descending spiral—do-si-la-sol-fa-mi-re-do—in the direction of the material creation, or to use Gurdjieff's term, involution, or an ascending spiral–do-re-mi-fa-sol-la-si-do—in the direction of greater consciousness, or evolution.

As described by Gurdjieff, a symbol has many intertwined meanings, which give it its explanatory power. As such it is a true reflection of many-faceted reality, while a step-by-step logical argument, although equally important, is a linear description of only one aspect of reality.

As an example of a multifaceted structure, consider the three brains of three-brained beings. Gurdjieff puts the center of gravity of the moving-instinctive brain in the spinal cord, of the emotional brain in the autonomic nervous system, and of the intellectual brain in the cerebrum. But, as discussed also elsewhere (*Does Man Have Three Brains?* In MiC, pp. 147–186), these three parts of the nervous system are extremely intertwined, and other subdivisions make equal sense. The front of the head brain (frontal lobes) can be seen as the intellect, and the back as concerned with the senses. In most people the left hemisphere is concerned with language and reasoning (more intellectual), and the right hemisphere with the emotional nuances of speech and music (more emotional). A third subdivision would have the cerebral cortex correspond to the intellect, the inner structures of the forebrain to the emotions, and the lower brainstem and spinal cord to the automatic functions of the body.

17

The enneagram in movement(s)

Based on existing records, one can conclude that Gurdjieff said relatively little specifically about the enneagram in his talks, despite the paramount importance he gave to it. There is a lecture on symbolism and the enneagram in Gurdjieff's Early Talks (GET, pp. 49–71), and some of the same material is reproduced in Ouspensky's chapter on symbols and the enneagram in *In Search of the Miraculous* (ISM, pp. 278–298). There is a little about the enneagram in *Views from the Real World* (VRW, pp. 214–221). There is no mention of it in Gurdjieff's own writings, although a fairly credible case has been made by Richard J. Defouw that the *structure* of Gurdjieff's first two books is based on the enneagram, and that, once deciphered, this sheds light on the enneagram's meanings.

On the other hand, the enneagram is a major component of Gurdjieff's movements, or sacred dances, which are a part of Gurdjieff's teaching and just as important as the verbal exposition of his ideas. Gurdjieff stated that experiencing the enneagram in movement was essential to understanding it:

> Much later—it was in the year 1922—when G. organized his Institute in France and when his pupils were studying dances and dervish exercises, G. showed them exercises connected with the "movement of the enneagram." On the floor of the hall where the exercises took place a large enneagram was drawn and the pupils who took part in the exercises stood on the spots marked by the numbers 1 to 9. Then they began to move in the direction of the numbers of the period [the hexagram] in a very interesting movement, turning round one another at the points of meeting, that is, at the points where the lines intersect in the enneagram.
>
> G. said at that time that exercises of moving according to the enneagram would occupy an important place in his ballet the "Struggle of the Magicians." And he said also that, without taking part in these exercises, without occupying some kind of place in them, it was almost impossible to understand the enneagram.

> "It is possible to experience the enneagram by movement," he said. "The rhythm itself of these movements would suggest the necessary ideas and maintain the necessary tension; without them it is not possible to feel what is most important." (ISM, pp. 294–295)

These statements fit well with Gurdjieff's characterization of human beings as three-brained, and his idea that all three brains must be active simultaneously and harmoniously for a fuller and more correct perception of reality to be possible. However, the three brains' ordinary relationships keep us in the prison of our habitual hypnotic sleep. The Gurdjieff movements are designed to evoke in us a new state of presence from which a truer perception is possible.

> Gurdjieff said: Every race, every period, every country, every profession has its own definite number of postures and movements.
>
> Every man has a definite repertory of roles. He has a role for every circumstance in his life.
>
> All our movements are automatic. Our thoughts and feelings are just as automatic.
>
> All our moving, thinking and emotional postures are connected. One cannot be changed without the other.
>
> Can man break this magic circle? Can he escape from his prison?
>
> Yes—if he suffers on account of this automatism and tries to oppose it.
>
> If he tries to find a new attitude in himself.
>
> But without help he will never be able to find it. Knowledge must be transmitted to him.
>
> New postures, proceeding from a different inner order, can be shown him.
>
> They express truths which are invisible in his ordinary state of consciousness.
>
> In practicing these postures, in passing from one to another, a man will try to understand their meaning and will gradually acquire direct knowledge of energy in movement.
>
> If he does, he will manifest another level of being.
>
> This is the significance of sacred dances.
>
> *(From the introduction to a film of movements made by Mme. de Salzmann, Gurdjieff's principal pupil, who preserved and developed Gurdjieff's movements after his death.)*

The movements are extremely multifaceted. Typically, multiple movements and rhythms are done simultaneously, requiring a level of attention that does not permit daydreaming and calls for a non-habitual sense of presence. The legs might be in one rhythm and the arms in another, the head in a third. One arm could be moving smoothly from one position to another, while the other moves abruptly. Yet the whole forms a pattern that is harmonious, reflecting the nested rhythms that prevail in all temporal organizations, from music to brain waves. At the same time, the different postures evoke different emotional states, in a logical succession.

> These Movements have a double aim. By requiring a quality of attention maintained on several parts at the same time, they help us to get out of the narrow circle of our automatism. And through a strict succession of attitudes, they lead us to a new possibility of thinking, feeling and action. (de Salzmann, p. 121–122)

Often there are words that are said, and/or a succession of inner sensing that is required.

The enneagram figures prominently in many of the movements. There are several like those that Ouspensky described, in which the participants move along the lines of the hexagram. In these, an interaction occurs at the points of the hexagram where two lines meet, so that, for instance, the moving dancer might be given the position to take to the next location by the dancer who stays in place. Alternatively, three people interact at the points of the hexagram, in a sequence of related movements, and then one moves on to the next location. These configurations symbolize the movements of energy required for the passage from one place to another in the enneagram, and the inner octaves that take place at each note (see below). The effect on the participants and the audience is not really describable in words, for multiple levels of perception are being evoked.

It is the interaction of multiple levels of vibration that comes through so clearly in the movements. This reflects Gurdjieff's idea of inner octaves and of cosmoses nested one within another. Each note of an octave representing a process contains a whole octave of notes within it, an octave on another level, as represented in the diagram on page 43. These nested rhythms need to be in the right relationship for processes to develop correctly. There are a number of somewhat obscure statements of Gurdjieff's that begin to make sense in terms of inner octaves, such as:

> The passage fa-mi in the cosmic octave is accomplished mechanically with the help of a special machine which makes it possible for fa, which enters it, to acquire by a series of *inner processes* the characteristics of sol standing above it, without changing its note, that is, to accumulate, as it were, the inner energy for passing independently into the next note, into mi. (ISM p. 291; italics mine)

The idea of inner octaves applies in many different situations. At the level of the overall economy, factories take in raw materials and produce certain products, while at the level within the factory, a whole variety of inner processes are taking place to make this possible. We eat food that is transformed into energy and the materials of our bodies, while at each stage of metabolism, a number of inner processes are occurring, and each step of these inner processes consists of another level of inner processes. At each level, there is a change of size and a change of speed. We eat three times a day, but the activities in our cells occur many times a second, and the molecular vibrations underlying them are much faster still.

Gurdjieff stated that the different centers function at different speeds, so they relate to each other as nested inner octaves. In particular, both the body and the emotions are much quicker than the ordinary mind, so that the idea that appropriate bodily tension and emotional attitude are necessary for understanding ideas such as those embodied in the enneagram makes perfect sense.

> At the same time, work on Movements allowed a direct experience of the laws governing the transformation of energy. This included the symbol of the Enneagram, which Gurdjieff said was almost impossible to understand without the feeling brought by participating in the Movements that are based on it. (de Salzmann, p. 122)

In relation to music, Gurdjieff described inner octaves as the slight variations of pitch that carry much of the emotional impact of music, as any singer who bends notes or violinist who uses vibrato is well aware. He describes 'objective art' as art in which these relationships are mathematically precise and result in predictable and reproducible effects.

> …without mathematical knowledge there can be no objective art. (VRW p. 180)

> Snake charmers' music in the East is an approach to objective music, of course very primitive. Very often it is simply one note which is long drawn

out, rising and falling only very little; but in this single note 'inner octaves' are going on all the time and melodies of 'inner octaves' which are inaudible to the ears but felt by the emotional center. (ISM p. 297)

It is these small variations of pitch and rhythm, as well as variations of timbre—overtone content—that give music much of its emotional impact. A completely steady rhythm, such as that produced by a rhythm machine, does not sound alive in the same way as a rhythm produced by a musician, and a steady tone lacks the quality of a musical note, with its subtle variations in pitch and timbre. These inner octaves speak to the emotional center and their proper production requires the emotional perception of the musician. Because they represent another cosmos nested within the larger cosmos of the music as written, they add another dimension to the music. The slightly different views that the two eyes have of the world leads to the direct perception of depth (see p. 58), and the slight variations of pitch, timbre, and rhythm in live music also create another dimension. In the movements, the participants try to be well synchronized with each other, but since each participant has a different body, there are slight variations there also, which add depth to the overall pattern formed by the class. It seems to be a general principle that anything alive has to contain inner octaves and consist of cosmoses nested within each other.

* * *

Far more numerous than movements done on a diagram of the enneagram on the floor are the 'multiplications.' In these movements, six files of dancers are designated as corresponding to the numbers 142857, and the participants change positions according to the multiplications by 2, 3, 4, 5, and 6. The successive arrangements of the files follow this pattern:

```
1 4 2 8 5 7
2 8 5 7 1 4
4 2 8 5 7 1
5 7 1 4 2 8
7 1 4 2 8 5
8 5 7 1 4 2
```

In the first multiplication, the two left files (as seen from the front) move to the right, and the other four files move to the left. Then, in many multiplications, everyone returns to the starting point. The second multiplication involves the first file moving to the right and the others to the left.

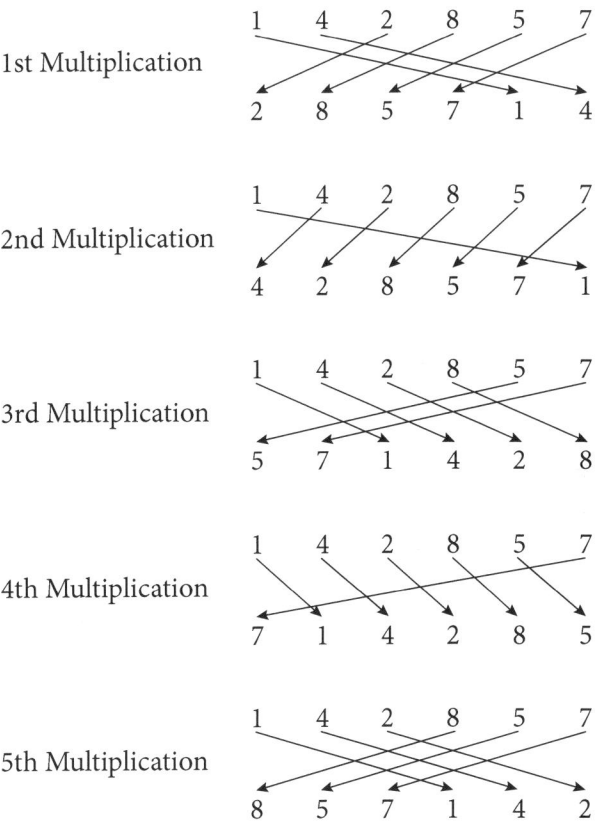

In the third multiplication, by 4, the two right files move to the left while the other four move to the right, and in the fourth, the rightmost file moves to the left, and the others to the right. The last multiplication is an exchange of three and three files. So the third and fourth multiplications are mirror images of the first and second, and could be regarded as representing opposite impulses, such as affirmation and negation, while the last is a reconciliation.

The movements are full of number patterns. Each participant's movements involve complex rhythmic patterns, and the movements of the class as a whole also involve patterns of numbers. Overall, the primacy of small number ratios in the workings of the universe is illustrated in many and varied ways.

* * *

In doing movements, one attempts to maintain an inner sensation of the body, and often one is also instructed to visualize each next position before taking it. Awareness of sensation is necessarily of the immediate past because of nerve conduction and processing delays. It has been estimated that it takes about half a second for a sensory stimulus to reach consciousness (Libet). Visualization of the next position is of the immediate future. In a mysterious way the combination of sensation and visualization thus frame and create the present. Two dimensions of time—ordinary time and the third dimension of anticipation and possibility—create the third, the eternal present. In one-dimensional ordinary external time alone, the present is an infinitesimal instant, between a past that is no longer and a future that does not yet exist. Time is inherently a property of the inner world. In Gurdjieff's words it is the "Ideally Unique Subjective Phenomenon." (BT p. 117)

18

More on the law of three

One of the two main laws governing the universe, according to Gurdjieff, is the law of three, which describes the interaction of three forces to produce any phenomenon. In *Beelzebub's Tales*, he explained the action of this law, called there 'triamazikamno', as follows:

> A new arising from the previously arisen through the "harnel-miatznel," the process of which is actualized thus: the higher blends with the lower in order together to actualize the middle, and thus to become either higher for the preceding lower or lower for the succeeding higher. (BT, p. 687)

He described the three forces in various ways, including

> the first, the 'affirming force,' or the 'pushing force' or simply the 'force plus'
> the second, the 'denying force' or the 'resisting force' or simply the 'force minus'
> the third, the 'reconciling force' or the 'equilibrating force' or the 'neutralizing force.' (BT, pp. 687–688)

As mentioned in chapter 4, all equations in mathematics and physics that describe interactions contain at least three terms, for a two-term equation is simply a definition or an equality. In a number of fundamental equations, the three terms seem to correspond quite well to Gurdjieff's three forces. In Newton's second law, force = mass × acceleration, force can be seen as pushing, mass as resisting, and acceleration as the result; however, it is more difficult to understand acceleration as reconciling or equilibrating—see the discussion of the difficulty of perceiving the third force in chapter 4.

Similarly, in Ohm's law, voltage = current × resistance, the voltage is the pushing force, the resistance the resisting force, and the current can be seen as the result. However, the roles of the three elements are not necessarily fixed, as one can also regard the current as the first force, the resistance as the second, and the voltage as the result. This interchangeability of terms suggests that the relationship of the three terms is more variable than we usually perceive, and that these equations describe a kind of resonance, rather than simply a causal interaction. This shift in perception requires a different understanding of time, various aspects of which are discussed throughout this book.

A more complex, or less obvious, three-element interaction is seen in the oscillations of a mass on a spring:

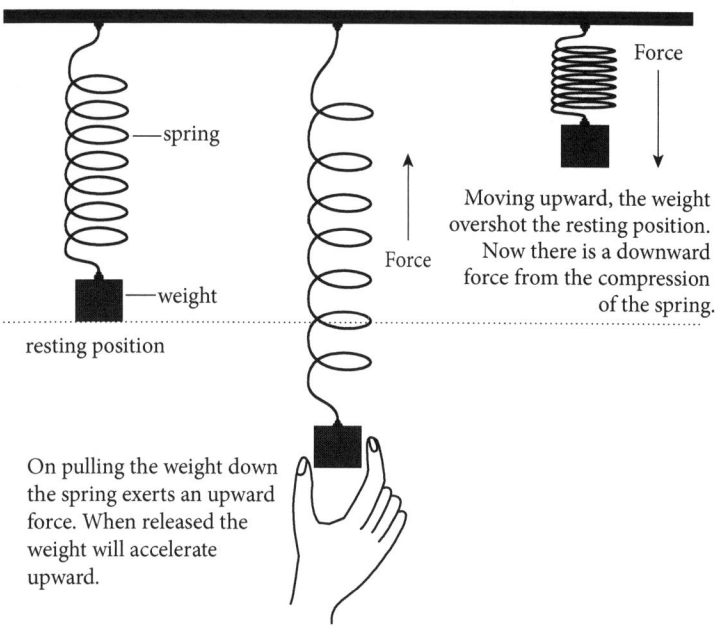

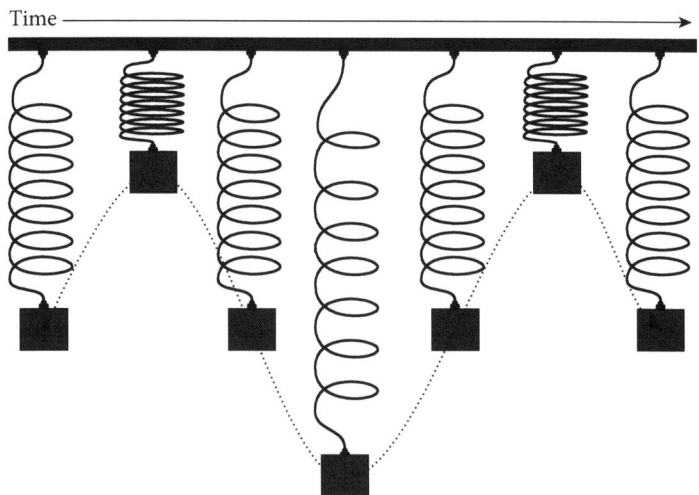

Once set in motion, the weight will oscillate sinusoidally. Eventually, because of friction, it will stop.

Time ⟶

Here the two opposing forces are the force resulting from the stretch or compression of the spring and the force associated with the accelerating mass. The two fluctuate in an opposite fashion: when the force of the stretch or compression of the spring is greatest, at the two extremes of the weight's trajectory, the weight is momentarily not moving; when it is moving the fastest, in the middle, the spring is at its resting position and not exerting any force. The result is an an oscillation, or vibration. The force of the spring is directly proportional to the distance of the mass from the resting position of the spring, while the force of the moving mass is proportional to its acceleration. Mathematically, the acceleration is the second derivative of position with respect to time; this terminology refers to the fact that acceleration is the instantaneous change of velocity with respect to time, and velocity in turn is the change of position with time. The resulting formula is:

$$m d^2x/dt^2 = -kx$$

m = mass   x = position   t = time   k = "stiffness" of the spring
d^2x/dt^2: second derivative of position with respect to time, or acceleration.

The reason for burdening the reader with this excursion into calculus is to point out that the two forces here are in a very real sense on different 'levels,' in that one is directly proportional to position while the other is proportional to the second derivative of position. This difference of level is often true of the three forces, although we typically do not perceive them that way. The importance of levels of forces, of materiality, and of cosmoses, in Gurdjieff's teaching is one of the things that sets it apart from the usual scientific perspective.

Gurdjieff stated that "knowledge begins with the teaching of the cosmoses" (ISM, p. 205). The cosmoses, such as those listed in the introduction, are not merely entities of different sizes, but each is regarded as a living being that is born, lives, and dies, and has a certain degree of consciousness. Further, Gurdjieff regarded them as related to each other as zero to infinity, in other words, as corresponding to dimensions. (As mentioned on page 58, the dimensional relationship of one cosmos to another in the real world [outside of abstract mathematics] is better described as *almost* zero to infinity.) What is the connection between scale, in the ordinary sense, and dimensionality, as applied to cosmoses?

In analyzing patterns of sound, such as speech or music, it is conventional to create a time-frequency plot, with time on the horizontal axis and frequency on the vertical. In other words, at any given time indicated on the x-axis, the frequencies of sound that are present are plotted on the y-axis. However, frequency is also time, or rather its reciprocal, being the reciprocal of the period, which is the amount of time for one cycle of the vibration.

Frequency = cycles/second   Period = seconds/cycle

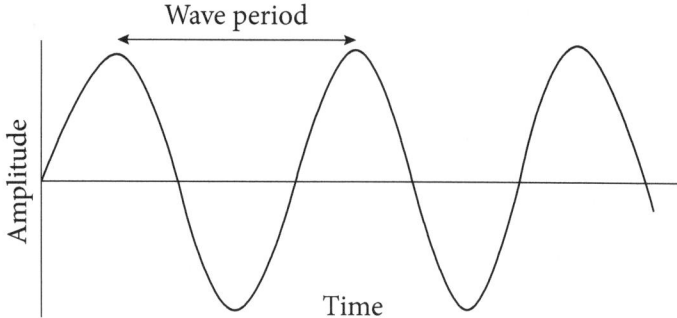

In addition, the time indicated on the horizontal axis is not instantaneous time, because it takes a certain amount of time for a frequency to be discernible. So the plot really is as follows: in a given time interval, say one second, what frequencies are present, such as 20 cycles/second, 50 cycles/second, etc.? Both axes refer to time, on different scales, with the vertical axis indicating the reciprocals of much shorter time intervals than those indicated directly on the horizontal axis. The two axes refer to two different time scales, represented as two different dimensions. The reciprocal relationship of the dimensions in such a plot is reminiscent of the reciprocal relationship between the outer cycle of the enneagram, representing seven steps in ordinary time, and the inner hexagram, representing the dimension of possibility by the fraction $1/7$.

Related to this is the fact that, according to quantum mechanics, it is not possible to simultaneously measure the time of an event and its energy with complete accuracy. In relativistic quantum mechanics, the energy of a particle is proportional to the frequency of its associated wave. So the time of an event and its energy are both time measurements, and the impossibility of measuring them both simultaneously with complete accuracy could suggest that they are related as two (approximately) perpendicular dimensions, and that a precise measurement along one dimension precludes precision along the other. The view of the cone as a triangle excludes its being seen as a circle. In measuring the exact time of a sound, one loses all information about its frequency, since determination of frequency requires an extended period of time.

The force of the stretched spring and the force of the moving weight run ahead of, or behind, each other, resulting in the oscillation. Similar timing relationships exist in electronic circuits that produce electromagnetic waves, such as radio waves, and the equations governing these circuits have the same form as those governing an oscillating weight on a spring. These considerations bring to mind the timing relationship between sensation and visualization in movements described at the end of chapter 17, and suggest that the present thus created has the nature of an oscillation. Electromagnetic oscillations in the brain seem to be intimately related to awareness, as discussed in chapter 15, and come in multiple nested frequencies. The half-second delay between a sensory stimulus and its conscious perception described by Libet may in part represent the time necessary for a sufficient sampling of frequencies.

*   *   *

A guitar string behaves in much the same way as a mass on a spring: plucking it stretches the string, which is fixed at both ends but has the capacity to stretch; the stretching acts like a spring and causes the string to accelerate back to its resting position. This acceleration causes the string to overshoot the resting position and stretch in the opposite direction, and so on. The vibration of the string sets air molecules in corresponding motion, and the air pressure fluctuations are perceived as sound.

When two sounds of different frequencies interact, another triad appears, because the interaction produces a third oscillation:

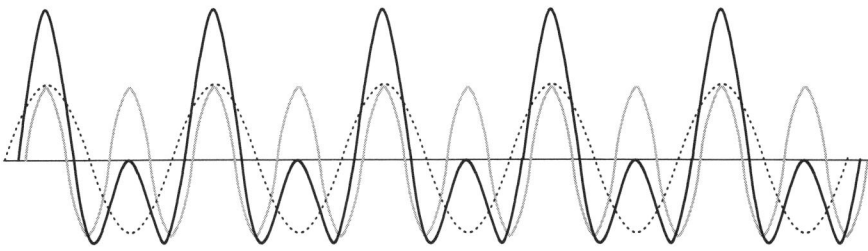

Solid grey wave has twice the frequency of the dotted wave. The resulting black wave is a point by point addition of the two component waves' amplitudes.

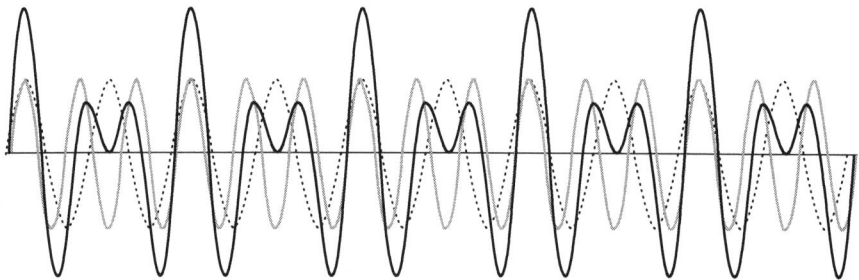

Wave pattern for a fifth in black.
Solid grey:dotted = 3:2

When the two component frequencies are only slightly different, the resultant wave fluctuates in amplitude more slowly: another vibration, called a beat pattern, of much lower frequency is produced. (See p. 57) This is a kind of broken symmetry, analogous to the creation of coarser and coarser worlds, with slower and slower vibrations, resulting from the broken symmetry of the law of seven.

Thus one can regard these various interactions as manifestations of the law of three.

* * *

Gurdjieff takes the law of three well beyond what we would call physics, regarding it as a universal law on all levels:

> ...the three-brained beings [of the planet Earth]...began to be aware of these three holy forces of the sacred Triamazikamno, which they named:
> the first, 'God the Father'
> the second, 'God the Son' and
> the third, 'God the Holy Ghost.'
> In various circumstances they expressed the hidden meaning of these forces and their longing to receive from them a beneficent effect for their own individuality by the following prayers:
> Sources of divine
> rejoicings, revolts, and sufferings,
> direct your actions upon us.
> or

> Holy Affirming,
> Holy Denying,
> Holy Reconciling,
> transubstantiate in me
> for my being.
>
> or
>
> > Holy God,
> > Holy the Firm,
> > Holy the Immortal,
> > have mercy on us.
>
> (BT, pp. 688–689)

Gurdjieff regarded the universe, and human beings, its servants—each one a miniature universe, made 'in the image of God'—as engaged in a perpetual struggle, which he described in various terms: between positive and negative, good and evil, active and passive, intention and resistance, conscious and mechanical, unity and dispersal, evolution and involution. This struggle maintains the universe by producing embodied consciousnesses.

"I am"—the sensation of self—the wish to be.

You begin as you can: first by the sensation of your body—I am there—the sensation of your individuality distinct from those around you. This is a fugitive sensation, it keeps disappearing. To maintain it, you have to watch it, which is an effort. You have to stay with it through "I wish" to make it last, to install it in yourself. You have to practice everywhere, at home, in the street, in the subway, in front of others. Especially in front of those who irritate us. When I am, there is no room for all that. I see the other. When I am not there, I do not see him, I cannot see him.

"I am" is the key. It is the only possibility of changing level. That's what is necessary for the passage. Without changing level, you can't understand more, know more. "I am" is a new attitude, an affirmation of self, a new attitude towards life, towards others. With "I am" you begin to feel the two natures, their terrific battles. It is the perpetual battle of the world, of the two Principles: the Good and the Bad; in other words of the positive, active principle, and of the negative. It's the most terrible of all wars. It is also in us. You must see it. Between these two principles which are our very nature, you must bring to birth an individual: the Man who is a measure, who has the power to make these two forces serve one Aim, the Man who acts for a Reason. (From notes reporting Gurdjieff's words at a meeting in the 1940s)

The last, apparently unfinished, chapter of Gurdjieff's last book, *Life Is Real Only Then, When "I Am"*, is entitled "The Outer and Inner World of Man." In it the task of human beings is described as being the bringing together of these two worlds, to create a 'third world of man'.

> Only he can have his own initiative for perceptions and manifestations in whose common presence there has been formed, in an independent and intentional manner, the totality of factors necessary for the functioning of this third world. (Gurdjieff, *Life is Real*, p. 173)

In this chapter, the outer world of man is described as "everything existing outside him, both what he can see and feel as well as what is invisible and intangible for him," and the inner world as "the automatic processes of his nature and the mechanical repercussions of these processes" (Gurdjieff, *Life is Real*, p. 173). Here the terms 'outer world' and 'inner world' are used differently than in other contexts, even almost in an opposite way, yet the same battle and possible result are being described. In yet another report of Gurdjieff's words, the coming together of two worlds is put in these terms:

> On another occasion, G. explained the idea of Moon from a wholly new direction. "Up to now we have talked about the Moon as the growing branch of the cosmos, as the end, or destination, of the Ray of Creation, which originates in the Absolute.
>
> "There is another level at which you must understand this idea: Given that man is the microcosm that replicates all that exists in the cosmos, this line from the Absolute to the Moon also exists in man. The representative of the Absolute in man is full consciousness, about which our knowledge is incomplete. We do know, however, that the effort to free oneself from identification creates a corresponding amount of free attention. The presence of free attention in a man is a second order representative of the Absolute: it is a foretaste of what he might eventually come to know as full consciousness.
>
> "The Moon-in-man is sensation. It is that broken part of the original consciousness of man, and it is that part towards which a man who wishes to work has a primary responsibility; for sensation in man is the growing part of his inner cosmos. The Ray of Creation inside man extends from free attention to sensation."
>
> In response to a question about the relationship between the growth of being and the growth of sensation, G. explained: "Just as the Moon in the sky requires vibrations from Earth for the growth of its atmosphere, sensation is the atmosphere of being. No growth of being will take place without

a corresponding growth in sensation.

"Of course, when we apply the term growth to sensation, we must understand that it refers to growth of the roots not the leaves, that is, sensation is not only of a man's skin, which we might think of as leaves, but of the entire inner structure, which includes the skeleton, muscles, and organs as well. In lifting his arm, everything on the other side of intention is sensation. A man must be able to radiate particles of free attention from the moment an intention enters his bloodstream and neurological system.

"The work on sensation is the infrastructure of being."

The many ways in which Gurdjieff describes the law of three, as it applies to the development of a 'permanent I,' or a second body, or a soul, in man, reflects the difficulty of understanding these ideas with the mind alone and the need for a three-brained perception. It is as if every verbal description is only partly correct, and so multiple versions are needed.

Gurdjieff also provides an anatomical description of the three brains, and a physiological description of the growth of a second body. He regarded the intellectual brain as centered in the head, the moving-instinctive brain in the spinal cord, and the emotional brain in the autonomic nervous system. Physiologically, the interplay of the three forces is described thus:

> In these nerve nodes [of the autonomic nervous system] scattered over the whole planetary body, all the results obtained from the affirming manifestations of their 'head brain' and denying manifestations of their 'spinal marrow' accumulate. And these results, once they are fixed in the nerve nodes scattered over the whole of their common presence, serve as the 'neutralizing principle' in the further process of affirmation and denial between the head brain and spinal marrow, in the same way as in the Megalocosmos the sum of all the results of the affirming manifestations of the Protocosmos and the various shades of denial of the newly arisen suns serve as the 'neutralizing force' in their further process of affirmation and denial.
>
> And so, like ourselves, the three-brained beings of the planet Earth are not only apparatuses—with the qualities of all three forces of the fundamental common-cosmic Triamazikamno—for transforming cosmic substances required for the Most Great Trogoautoegocrat, but also have the possibility, while absorbing these substances coming from three independent sources, of assimilating, in addition to the substances indispensable for their own existence, certain substances destined for the coating and perfecting of their own higher being-bodies. (BT, p. 714)

It is clear, from the many varied kinds of descriptions and analogies he used, that Gurdjieff believed that the growth of being in humans, resulting ultimately in a 'permanent I' and the first stage of the soul, is to a considerable extent a function of the growth or maturing of the emotional part, which part is the most immature in many people.

In equating the emotional brain with the autonomic nervous system, Gurdjieff is indicating that the potential functions of this part of the nervous system extend well beyond those known to ordinary physiology, of regulating the energies of the body in an automatic way. He stated that the 'subconsciousness', meaning this emotional brain, should become our true consciousness, and that objective conscience is buried there; were objective conscience to become active in our waking lives, we would begin to approach the state in which the mutually aware consciousnesses of three-brained beings could serve the consciousness of the universe.

In the enneagram, the law of three appears in multiple ways. At each point in the development of the octave three forces act, as three forces are necessary to effect any transformation. This is clearly depicted in one of the movements based on the enneagram, in which three people interact in a series of moving patterns at each point of the hexagram, before sending one person to the next point of the sequence 142857.

The triangle also represents the three forces, as depicted in one of Gurdjieff's own diagrams(ISM, p. 293)

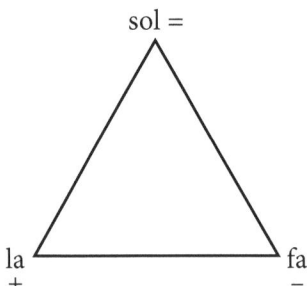

and in another enneagram-based movement in which three people, seated at the three points of the triangle, repeat in turn the words affirmation, negation, and reconciliation, while the other participants move along the hexagram in the order 142857.

The placement of the three notes in the above diagram may seem odd, as sol, at the top, is between fa and la in frequency, rather than above or below them, as one might expect from its location. However, this placement depicts the struggle discussed above, between the active and passive forces, represented by the left and right sides of the enneagram, resulting in the formation of an individual with its own "I," represented by sol, at the top. This configuration also represents a return to the source, from the multiplicity represented by the base of the triangle.

In addition, the sequences 142 and 857 form two triads, which aid the passages from mi to fa and from sol to la, as depicted in the enneagram.

* * *

One way to represent the law of three is with a braid. In braiding hair or other materials first the right (or left) strand is brought over the middle one, and then the left (or right) is brought over the new middle, and so on in an alternating pattern. If one starts with the right strand, regarded as the higher, and brings it over the middle one, regarded as the lower, the new middle is the result, and becomes lower for the next higher, the left strand. If one labels the strands, arbitrarily, Red, Green, and Blue, one gets a sequence of configurations of the strands that returns to the original arrangement after six iterations:

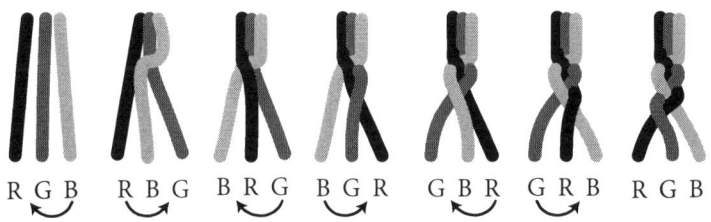

RGB, RBG, BRG, BGR, GBR, GRB, RGB

One can place this sequence on the enneagram, following the numbers 142857:

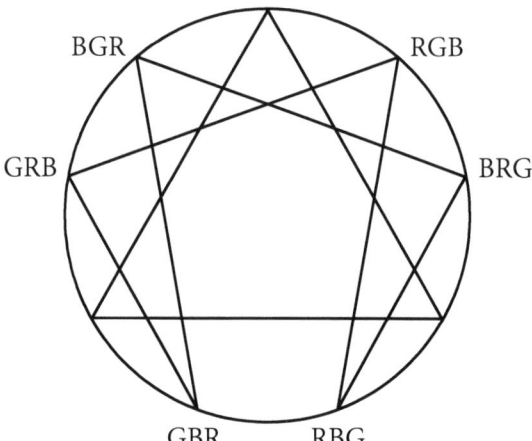

Interestingly the triads of colors on the same level are mirror images: RGB at point 1 is the mirror image of BGR at point 8. Similarly, BRG at 2 mirrors GRB at 7, and RBG at 4 mirrors GBR at 5. In parallel, each pair of numbers—1 and 8, 2 and 7, and 4 and 5—adds up to 9.

If one starts braiding with the left strand, the sequence is RGB, GRB, GBR, BGR, BRG, RBG, RGB. This is the reverse of the order that results from starting with the right strand, and can also be obtained from the enneagram of the first sequence by moving backwards, along 758241.

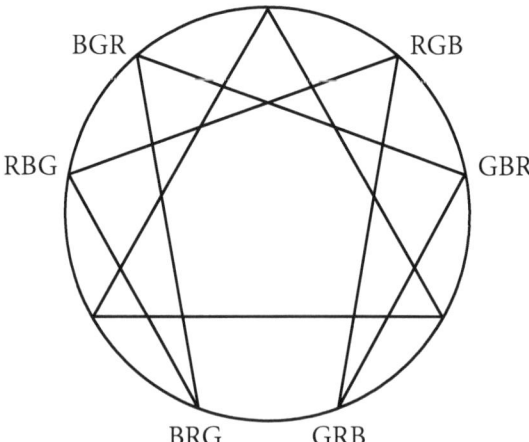

Braids form a mathematical group under combination: putting two braids end to end creates another braid; there is an identity, the untangled strands, and every braid has an inverse, which will unbraid it and produce the identity.

Since the numbers 124578 form a group under multiplication, one wonders whether the two symmetries, the mirror symmetry of the braids on the enneagram. and the fact that the corresponding numbers all add up to 9, are related, but I am not enough of a mathematician to prove this (or disprove it).

If one places the letters RGB vertically on a möbius strip and follows their positions as one goes around the strip, they will be in an upside-down configuration when one has gone around once, and it takes two revolutions around the strip to get back to the original orientation. This is intriguingly similar to the fact that spin ½ fermions need to "turn around" twice to return to the same state.

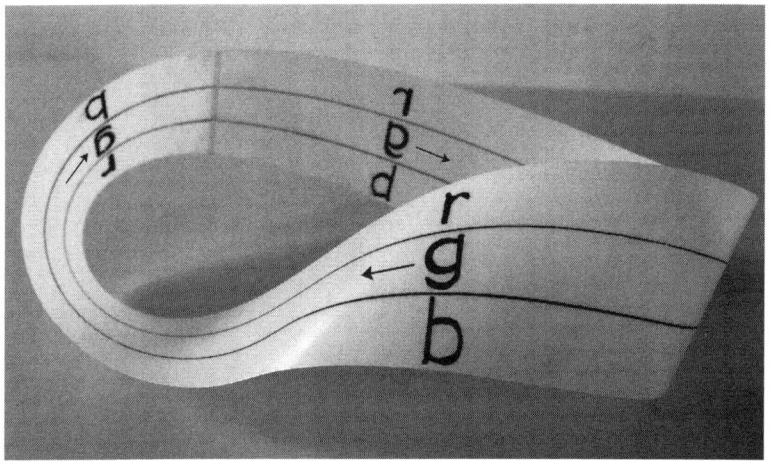

19

More on the law of seven

If the just-intonation major scale represents a universal law governing processes, as Gurdjieff claims, one would think that there must be some sort of physical and mathematical inevitability about it. One approach to this is to ask to what extent this scale arises "naturally." As described in chapter 8, the first four octaves of the overtone series of a vibrating string contain many of the notes of the just scale, but several of the notes of the scale are not in the overtone series, and conversely, several of the overtones are not in the just scale. Is there justification for these discrepancies?

The first four octaves of the overtone series, as described in chapter 8, are reproduced below, for a fundamental tone of 100 Hz:

100	do
200	do
300	sol ($3/2$ of second do)
400	do
500	mi ($5/4$ of third do)
600	sol ($3/2$ of third do)
700	not a note of the just scale
800	do
900	re ($9/8$ of fourth do)
1000	mi ($5/4$ of fourth do)
1100	not in the just scale
1200	sol ($3/2$ of fourth do)
1300	not in the just scale
1400	not in the just scale
1500	si ($15/8$ of fourth do)
1600	do

The notes of the just scale, with the ratios of their vibration rates compared to the first note, are:

1	9/8	5/4	4/3	3/2	5/3	15/8	2
do	re	mi	fa	sol	la	si	do

So, on the one hand, fa and la are notes of the just scale not present in the overtone series, because of the 3 in their denominators, and on the other, 700 (7/4), 1100 (11/8), and 1300 (13/8) are not in the just scale despite their presence in the overtone series (14/8 is the same as 7/4). It would seem sensible to include fa and la in the scale as they represent small number ratios and sound very harmonious in music. Their position in the enneagram is also intriguing, as they follow the two places where outside influences enter into a process.

It is perhaps interesting in this regard that in going from do-mi to do-fa musically there is a change in the quality of the harmony: do-mi, the major third, has a sweet quality, whereas do-fa, the fourth, is a more austere—one could even say classic as opposed to romantic—harmony. The transition from do-sol, the fifth, to do-la, the major sixth, is the reverse, from a more austere harmony back to a sweeter one.

On the other hand, the exclusion of 7/4, 11/8, and 13/8 from the scale could, in some mysterious way, be related to their numerators, as these are the only numerators in the overtone series that do not divide 360. As noted in chapter 6, the numbers 2, 3, 4, 5, 6, 8, 9, 10, 12, and 15 all evenly divide 360, recognized as a special number since antiquity.

The two-note combinations that sound harmonious to the ear are those involving small number ratios: 2:1, 3:2, 4:3, 5:3, and 5:4. The minor third, 6:5, is at the borderline of sounding harmonious, which is perhaps why it has a sad quality. 9:8 and 15:8 do not sound harmonious but are needed in the scale as stepping stones to the other notes: 9:8—re—leads to mi, which is harmonious with do, or back to do, and 15/8—si—leads very strongly to the octave do. Of course, strict and constant harmony is not only what music is about, rather there is a constant interplay between dissonance and harmony; otherwise, music could not move and progress emotionally. Movement from one harmony to another without intervening dissonances would not reflect the workings of the world. For new chemical compounds to be formed, others have to first be taken apart. 7/4, 11/8 and 13/8 do not make harmonious sounds with the base note of the scale but would seem to be unnecessary in the middle of the scale; the semitones between whole notes can provide any dissonant elements necessary for movement of the music.

In some musics, smaller intervals than semitones are used between the main notes of the scale—quarter tones or even smaller intervals. In fact, in

Gurdjieff's first exposition of his ideas about the just scale, he described two 'semitones' between most of the main notes and only one between mi-fa and si-do (ISM, p. 126). The significance of this is unclear to me.

These somewhat tentative musings about musical harmony nevertheless reflect the fact that our brains resonate to some musical ratios in a pleasing way and not to others. Why the complex electromagnetic vibrating structures that are our nervous systems have these preferences is not clear, but perhaps these preferences relate to more general physical laws of vibrations and resonance. Further, perhaps movement in music reflects general laws of how processes develop over time. This is Gurdjieff's claim.

* * *

The ancient Pythagoreans were well aware of the behavior of powers of 2 and powers of 3, and the difference between $(3/2)^{12}$ and 2^7, as described in chapter 8, is known as the Pythagorean comma. They were also very interested in different kinds of proportions between numbers, or *means*. There are three kinds of mathematical means. The *arithmetic* mean between two numbers is the number that is equidistant from the two: 4 is the arithmetic mean between 2 and 6. The *geometric* mean is multiplicative: 2 is twice 1, and 4 is twice 2, so 2 is the geometric mean between 1 and 4. The *harmonic* mean is such that the ratio of the two extreme numbers is the same as the ratio of their differences from the middle number: the harmonic mean between 6 and 12 is 8 because 12 is twice 6, and 4 (the difference between 8 and 12) is twice 2 (the difference between 8 and 6). The harmonic mean is called that because it is the mean involved in the harmonic series, which describes the different ways a string fixed at both ends can vibrate: in halves, thirds, quarters, etc.

The harmonic series: 1 $1/2$ $1/3$ $1/4$ $1/5$ $1/6$ $1/7$ $1/8$ etc.

Thus 1/2 is the harmonic mean between 1 and 1/3: the three numbers can be represented as 6/6, 3/6 and 2/6, and 6 is three times 2, just as 3 (the difference between 6 and 3) is three times 1 (the difference between 3 and 2). Any adjacent three numbers in the harmonic series have this relationship.

Taking all these numerical facts into account, one can arrive at a different way of generating a scale that resembles the just scale with regard to the relationships of the intervals (Michell):

The Pythagoreans arranged the powers of 2 and of 3 in a configuration resembling the capital Greek letter lambda:

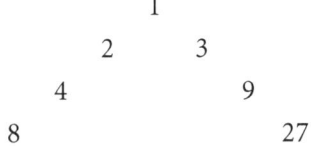

Thus the numbers along each leg of the lambda are in geometric proportion.

Between each pair of numbers, they placed the other two means: arithmetic and harmonic. To do this, for the left leg using whole numbers, everything must be multiplied by 6:

$$\begin{matrix} 6 \\ 12 \\ 24 \\ 48 \end{matrix}$$

Then the means between each pair of numbers are

6, <u>8</u>, **9**, 12, <u>16</u>, **18**, 24, <u>32</u>, **36**, 48.

Here the harmonic means are underlined and the arithmetic means are in bold.

The ratios between adjacent numbers of this series are 4:3, 9:8, 4:3, 4:3, 9:8, 4:3, 4:3, 9:8, 4:3. If one then factors 4:3: 4:3 = 9:8 × 9:8 × 256:243, one gets the following series of intervals, repeated three times:

do	re	mi	fa	sol	la	si	do
	9:8	9:8	256:243	9:8	9:8	9:8	256:243

Just as in the just major scale, there are smaller intervals between mi and fa and between si and the octave do. It is intriguing that this convoluted method involving proportions results in a scale that has the same characteristics as the just major scale.

20
Conclusion

Gurdjieff's musical model of universal laws and the patterns of the enneagram reflect the importance and ubiquity of vibrations. He stated:

> In order to understand the meaning of this law [the law of seven] it is necessary to regard the universe as *consisting of vibrations*. These vibrations proceed in all kinds, aspects, and densities of the matter which constitutes the universe, from the finest to the coarsest; they issue from various sources and proceed in various directions, crossing one another, colliding, strengthening, weakening, arresting one another, and so on. (ISM, p. 122)

I don't think any scientist would disagree with this statement, provided we include the vibrations of probability amplitudes that rule the quantum world.

Increasingly, it is thought that the functioning of the brain also relies on vibrations, or oscillations, the preferred term of neuroscientists. These vibrations consist of relatively synchronized fluctuations of the membrane potentials of many cells and are reflected in the brain waves recorded by electroencephalography or magnetoencephalography. One of the features of these vibrations is that they are "nested" one within another, or 'phase amplitude coupled' (Lakatos). This means that a slower rhythm will "contain" a number of vibrations of a faster rhythm, and the amplitude of the faster waves will fluctuate with the phase of the slower ones—at the peak of the slower rhythm, the faster waves might have their greatest amplitude, and at the trough of the slower rhythm, the faster waves their lowest amplitude.

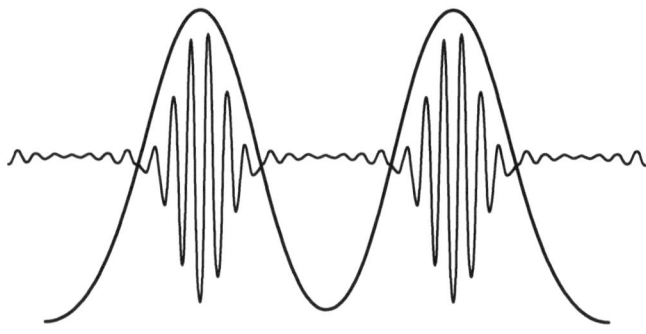

This picture illustrates the concept, but of course in real brain recordings, the correlations between rhythms are constantly changing and the patterns are messier, often requiring statistical analysis to become apparent.

In the experiments related to the 'cocktail party problem,' described on p. 92, the subjects' brain waves reflected the speech rhythms of the attended to speaker. It is likely that within this rhythmic envelope, faster rhythms relating to the actual content of the story were nested. The slower rhythm enables the faster ones to be effectively captured and attended. To push a child on a swing, the rhythm of pushing has to match the natural rhythm of the swing; in everything that unfolds in time, rhythms have to be in harmonious relationships for things to work properly.

It is the same with music, in which at every level rhythms are included one within another. In a four-beat measure, there may be four quarter notes, or eight eighth notes, or a half note and two quarter notes, or four triplets; in a major chord, the third vibrates at 5/4 the rate of the tonic and the fifth at 3/2 the rate, while the overtones create a multitude of further nested relationships.

So, if the world is made of vibrations, these kinds of patterns are its currency. Furthermore, entities that persist in time must consist of partially stable vibrational patterns. Vibrations denote order in time, and complex enduring entities like organisms must consist of extremely complex partially stable vibrational patterns.

The matrices that regulate what is possible in the world of atoms and subatomic particles are mathematical patterns. The probability amplitude vibrational patterns of these atoms and particles are more complex and multidimensional than the vibrational modes of a guitar string. But the principles are the same.

As noted in chapter 7, a torus, or donut shape, can represent two perpendicular dimensions, or cosmoses, one nested within the other. The smaller circle corresponds to the faster vibrations in one cosmos, the larger circle to the slower ones in the other. This also corresponds to the idea of inner octaves. However a generic torus has no requirement that the two circles' circumferences be in an integer relationship, as is the case with musical harmonies and harmonics. So, to represent two adjacent nested cosmoses more correctly, perhaps a 'quantized' torus is needed.

N-tori, or hypertori, involving more than two perpendicular circles or dimensions, are topological objects studied in higher mathematics. Perhaps a quantized n-torus with seven dimensions (7-torus) could represent the relationships and interactions of the cosmoses in the universe.

* * *

The extent to which the patterns found in the enneagram are correlated with the laws of physics is largely unexplored, although intriguing parallels have been looked into in this book. Nevertheless, it seems clear that mathematical patterns determine what is possible in the world, and the enneagram captures quite a few of the prominent patterns involving the relationships of the first 10 numbers.

The enneagram, as interpreted here and elsewhere (Bennett, Blake) suggests possible developments in our understanding of universal laws, especially with regard to the nature of time and the role of consciousness, both of which have remained stubborn enigmas for scientists so far (*A Question of Time—The Ultimate Paradox*, Scientific American Editors, 2012). The inclusion of time and consciousness, and of both the inner world of awareness and outer world of phenomena, in the enneagram suggests a larger view than is currently encompassed by modern science, as befits a "universal symbol," while the predictive precision of modern science is missing, so far, in the interpretations of the enneagram. This reflects the different kinds of precision exemplified by a symbol as opposed to a formula (see chapter 3). One would hope for a science that could include both. Perhaps those with more training in mathematics and physics than I could take these explorations further.

I am well aware of the fact that, given enough bits of information, one can make connections between almost anything and almost anything else. For instance, if one took all the measurements of the Empire State Building, such as the sizes of the windows and doors, the distances between this and that, the numbers of screws of given sizes, etc., etc., one could possibly, by selectively picking some of these numbers, make a connection between the measurements of the building and, say, the orbits of the planets, which would be entirely spurious. Humans are always looking for explanatory patterns, and some that we come up with seem arbitrary and unrelated to the real workings of the world. On the other hand, it would seem that we look for patterns because we are tuned to the real patterns that govern things. In this book, I have elaborated on a large number of connections between various patterns of numbers, without tying everything all together into a complete theory, or one that currently leads to precise predictions of experimental results. I strongly suspect that most of the patterns found in the enneagram reflect real correlations with the actual laws that govern things, and are not spurious, but the search for truth is always filled with pitfalls.

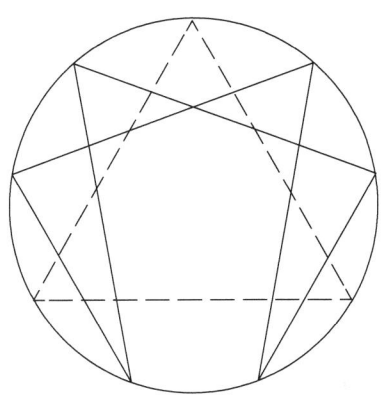

Glossary

Words invented by Gurdjieff that appear in BT, explained on their first appearance in the quotes used in this text, but not necessarily thereafter.

Harnel-aoot: the fifth interval in the scale, between sol and la.

Heptaparaparshinokh: the law of seven.

Heropass: time.

Higher being-bodies: stages of the soul; souls.

Megalocosmos: the universe.

Mdnel-in: special name given to the 3rd and 7th stopinders, or intervals, in the scale.

Microcosmos: the cell.

Okidanokh: akin to electromagnetism.

Protocosmos: first cosmos, or God.

Stopinder: an interval in the musical scale.

Triamazikamno: the law of three.

Trogoautoegocrat: reciprocal maintenance of everything in the universe; all processes need help from outside themselves.

References

Archive.org. Full text of "Orage Gurdjieff Meeting Notes," p. 67. There is uncertainty as to the origin of this text. This portion of the meeting notes is entitled "Notes taken by Blanche B. Grant on A.R. Orage's group talks." In a book by Louise Welch, *Orage with Gurdjieff in America* (Boston, London, Melbourne and Henley: Routledge & Kegan Paul, 1982), pp. 79–80, a similar partial text is attributed to Orage, one of Gurdjieff's principle pupils. Many people are quite sure that the original thoughts came from Gurdjieff.

Baez J. My Favorite Numbers. The Rankin Lectures, presented at University of Glasgow Lecture Series, September 15-19, 2008. Available at YouTube.com.

Beke GL. *Digging Up the Dog, The Greek Roots of Gurdjieff's Esoteric Ideas.* New York, NY: Indications Press; 2005.

Bennett JG. *Enneagram Studies.* York Beach, ME: Samuel Weiser, Inc; 1983.

Blake AGE. *The Intelligent Enneagram.* Boston, MA, & London, England: Shambala; 1996.

Bouanchaud B. *The Essence of Yoga: Reflections on the Yoga Sutras of Patanjali.* Desneux R, translator. Portland, OR: Rudra Press; 1997.

de Salzmann J. *The Reality of Being—The Fourth Way of Gurdjieff.* Boston, MA, & London, England: Shambala; 2010.

Defouw RJ. *The Enneagram in the Writings of Gurdjieff.* Indianapolis, IN: Dog Ear Publishing; 2011.

Eliot C. Net: Indira's net, Hindu mythology. Quoted by: University of Kent Web site. References, definitions of "Indra's net": net, network, web metaphor. https://www.cs.kent.ac.uk/people/staff/saf/networks/networking-networkers/indras-net.html. Accessed September 2, 2017.

Gray T, Manogue CA. *The Geometry of the Octonions.* Hackensack, NJ: World Scientific Publishing Co Pte Ltd; 2015.

Gurdjieff GI. *Life Is Real Only Then, When "I Am," The Third Series of All and Everything.* New York, NY: EP Dutton & Co, Inc, for Triangle Editions, Inc; 1975.

Guthrie KS, Fideler D. *The Pythagorean Sourcebook and Library.* Grand Rapids, MI: Phanes Press; 1987.

Haber HE, Kane GL. Is nature supersymmetric? *Sci Am.*, June 1986:52.

Huntley HE. *The Divine Proportion: A Study in Mathematical Beauty.* New York, NY: Dover Publications, Inc; 1970:160–163.

Jayaram, V. The Triple Gunas, Sattva, Rajas, and Tamas. Hindu Web site. http://hinduwebsite.com. Accessed September 2, 2017.

Katha Upanishad, edited by the author based on multiple translations.

Lakatos P, Shah AS, Knuth KH, Ulbert I, Karmos G, Schroeder CE. An oscillatory hierarchy controlling neuronal excitability and stimulus processing in the auditory cortex. *J Neurophysiol.* 2005;94(3):1904–1911.

Libet B, Pearl DK, Morledge DE, Gleason CA, Hosobuchi Y, Barbaro NM. Control of the transition from sensory detection to sensory awareness in man by the duration of a thalamic stimulus: the cerebral time-on factor. *Brain.* 1991;114:1731–1757.

Livio M. *The Golden Ratio: The Story of Phi, the World's Most Astonishing Number.* New York, NY: Broadway Books; 2002.

Michell J. *The Dimensions of Paradise—The Proportions and Symbolic Numbers in Ancient Cosmology.* New York, NY: Harper & Row, Publishers; 1988.

Sargeant W. *The Bhagavad Gita.* Albany, NY: State University of New York Press; 2009.

Schoedinger I. The Oneness of Mind. In: Wilber K, ed. *Quantum Questions: Mystical Writings of the World's Great Physicists.* Boston, MA, and London, England: Shambala New Science Library; 1985.

Strohmeier J, Westbrook P. *Divine Harmony: The Life and Teachings of Pythagoras.* Berkeley, CA: Berkeley Hill Books; 1999.

Taylor T. *The Theoretic Arithmetic of the Pythagoreans.* York Beach, ME: Samuel Weiser, Inc; 1983.

Young AM. *The Reflexive Universe: Evolution of Consciousness.* Mill Valley, CA: Robert Briggs Associates; 1976:259–282.

Zion Golumbic EM, Ding N, Bickel S, et al. Mechanisms underlying the selective neuronal tracking of attended speech at a "cocktail party." *Neuron.* 2013; 77:1–12.